MW01295765

ARCHANGEL:

CIA's SUPERSONIC A-12
RECONNAISSANCE AIRCRAFT

DAVID ROBARGE
CIA CHIEF HISTORIAN

CENTRAL INTELLIGENCE AGENCY
WASHINGTON, D.C.
Second Edition
2012

CENTER FOR THE
STUDY OF INTELLIGENCE

The Center for the Study of Intelligence (CSI) was founded in 1974 in response to Director of Central Intelligence James Schlesinger's desire to create within CIA an organization that could "think through the functions of intelligence and bring the best intellects available to bear on intelligence problems." The Center, comprising both professional historians and experienced practitioners, attempts to document lessons learned from past activities, explore the needs and expectations of intelligence consumers, and stimulate serious debate on current and future intelligence challenges.

To support these efforts, CSI publishes Studies in Intelligence and books and monographs addressing historical, operational, doctrinal, and theoretical aspects of the intelligence profession. It also administers the CIA Museum and maintains the CIA Library's Historical Intelligence Collection.

Comments and questions may be addressed to:

Center for the Study of Intelligence
Central Intelligence Agency
Washington, DC 20505

Second Edition, January 2012

Table of Contents

FOREWORD

This history of the A-12 reconnaissance aircraft was occasioned by CIA's acquisition on loan from the Air Force of the eighth A-12 in the production series of 15 in September 2007. Known as Article 128, the aircraft is on display at the Agency's Headquarters compound in Langley, Virginia. This history is intended to provide an accessible overview of the A-12's development and use as an intelligence collector.

Writing this story was a fascinating challenge because I am not an aviation historian and have never flown any kind of aircraft. Accordingly, I have tried to make the narrative informative to lay readers like myself, while retaining enough technical detail to satisfy those more knowledgeable about aeronautics and engineering. I have drawn on the sources listed in the bibliography and the extensive files on the A-12 program in CIA Archives. Hundreds of those documents were declassified and released to the public in conjunction with the dedication of Article 128 in September 2007 as part of the Agency's 60th anniversary commemoration. I have limited citations to specific documentary references and direct quotes from published works. When discrepancies arose among the sources regarding dates and other details, I have relied on the official records.

For their contributions to the substance and production of this work and to the documentary release, I would like to thank my colleagues on the CIA History Staff and at the Center for the Study of Intelligence, the information review officers in the Directorate of Science and Technology, designers and cartographers in the Directorate of Intelligence, and publication personnel at Imaging and Publishing Support. I also am grateful for historical material provided by the Lockheed Martin Corporation and the A-12 program veterans, the Roadrunners.

David Robarge CIA Chief Historian

January 2012

FROM DRAWING BOARD TO FACTORY FLOOR

The Central Intelligence Agency was created in 1947 principally to provide US leaders with strategic warning of attack by the Soviet Union. The Agency's main mission during its first decade and a half was to deploy its collection and analytic assets to detect and preempt a nuclear Pearl Harbor. No other intelligence question had greater implications for the national interests of the United States, and its very survival, than determining what kinds of strategic weapons, and how many of them, the Soviet Union had, and how it intended to use them. With the USSR proving to be an extremely hard target for traditional espionage operations, the United States had to turn to technical collection to peer beyond the Iron Curtain.

In 1954, CIA retained the Lockheed Corporation to build the U-2 reconnaissance aircraft. Essentially a jet-powered glider, the U-2 could fly at the unprecedented height of 70,000 feet, beyond the range of Soviet fighters and missiles, and take detailed photographs of Soviet Bloc military facilities. The aircraft was ready for operations in June 1956. At the time, CIA project officers had estimated that the U-2 would

3

be able to fly safely over the Soviet Union for two years at most before it became vulnerable to Soviet air defenses. The Soviets tracked the U-2 from its first mission, however. The estimate had proven too optimistic, especially after initial efforts to mask the U-2's radar image proved ineffective. A more radical solution was needed, an entirely new aircraft.

Genesis of the A-12 Design - GUSTO

After a year of discussion with aviation companies beginning in the late summer of 1956, CIA focused its attention on building a jet that could fly at extremely high speeds and altitudes while incorporating state-of-the-art techniques in radar absorption or deflection. This effort was codenamed GUSTO. In the fall of 1957, U-2 project manager Richard Bissell established an advisory committee to help select a design for the U-2's successor. Chaired by Polaroid chief executive Edwin Land, the committee met seven times between November 1957 and August 1959. Designers from several aircraft manufacturers and senior officials from the Navy and Air Force attended some of the meetings. The two most prominent firms involved in the process were Lockheed, which already was investigating designs for the U-2's replacement, and Convair, which was building a supersonic bomber for the Air Force, the B-58 Hustler.

Lockheed's chief engineer, Clarence "Kelly" Johnson, said that "It makes no sense to just take this one or two steps ahead, because we'd be buying only a couple of years before the Russians would be able to nail us again....I want us to come up with an airplane that can rule the skies for a decade or more." On 21 April 1958, Lockheed's Advanced Development Projects component, jokingly nicknamed the "Skunk Works" after the backwoods moonshine still in the comic strip *Li'l Abner* and already responsible for so many cutting-edge aviation achievements, began designing an aircraft that would cruise at Mach 3.0 at altitudes above 90,000 feet. "The higher and faster we fly, the harder it will be to spot us, much less stop us," Johnson asserted (Rich and Janos, 193). On 23 July 1958, Johnson presented his concept to Land's committee, which expressed interest in the approach. By September, the Skunk Works had studied various configurations called "Archangel-1," "Archangel-2," and so forth, a carryover from the original moniker of "Angel" given to the U-2 during its development. The nomenclature soon became simply "A-1," "A-2," etc.

That same month, the Land committee met to review all concepts proposed so far. It rejected two Lockheed designs, for a tailless subsonic aircraft and a new supersonic model, the A-2, and approved continuation of Convair's work on a Mach 4 "parasite" aircraft called the FISH that would be launched from beneath a modified Hustler called the B-58B. In November, Lockheed submitted its design for the A-3, which was much smaller and lighter than the previous concepts. Upon receiving recommendations from the committee, President Dwight Eisenhower in December approved funding for further R&D on the U-2's successor. Bissell asked Convair and Lockheed to turn in detailed proposals. Both firms did by the summer of 1959.

Lockheed worked on seven more designs in the ensuing months. None of them met all the requirements for speed, altitude, range, and radar cross section (RCS). Johnson decided not to sacrifice aerodynamic capability to achieve a smaller RCS.

In March 1959, Lockheed developed a design for the A-11. It would have a top speed of Mach 3.2, a range of 3,200 miles, and an altitude of 90,000 feet, and could be ready by January 1961. The A-11's main drawback was that it would be more detectable than Convair's much smaller FISH. That manned, ramjet-powered vehicle was designed to fly at Mach 4.2 at 90,000 feet with a range of 3,900 miles, and also could be ready by January 1961. Three substantial uncertainties beset the FISH, however: the unproven technology of ramjet engines; the unavailability of the B-58B that would fly fast enough to launch it; and the possibility that the B-58B could not reach the necessary speed, or that if it did, the FISH could not operate under post-launch conditions. The Air Force's cancellation of the B-58B project in June 1959 took the FISH out of the running, but the Land committee also rejected the A-11 because its RCS was still too large. The competition continued.

Convair and Lockheed completed new proposals in August 1959. Convair's entry, known as the KINGFISH, was a ground-launched, single-pilot jet with two Pratt & Whitney J58 engines, the most powerful available, and a small RCS. Lockheed's design, the A-12, also would use the J58 engines. It would reach Mach 3.2 at up to 97,600 feet and have a range of around 4,600 miles. To save weight, Johnson decided not to construct the aircraft out of steel. Because standard lightweight metals such as aluminum could not withstand the heat generated at Mach 3 speeds, Johnson chose a titanium alloy. The A-12's design incorporated a continuously curving airframe, a forebody with tightly slanted edges called chines, engine housings (nacelles) located mid-wing, canted rudders, and nonmetallic parts to decrease the RCS. A cesium fuel additive would reduce the radar detectability of the afterburner plume.

The two firms submitted their designs to a selection panel with members from the Department of Defense, the Air Force, and CIA on 20 August 1959. The A-12's specifications were slightly better than the KINGFISH's, and its projected cost was significantly less. Convair's design had the smaller RCS, however, and CIA's representatives initially favored it for that reason. The companies' respective track records proved decisive. Convair's work on the B-58 had been plagued with delays and cost overruns, whereas Lockheed had produced the U-2 on time and under budget. In addition, it had experience running a "black" project. On 28 August, Johnson wrote in his project log,

> *Saw Mr. Bissell alone. He told me that we had the project and that Convair is out of the picture. They [CIA] accept our conditions (1) of the basic arrangement of the A-12 and (2) that our method of doing business will be identical to that of the U-2. He agreed very firmly to this latter condition and said*

that unless it was done this way he wanted nothing to do with the project either.(Johnson, Archangel Log, 7)

Much of the eventual success of the OXCART program can be attributed to CIA and Lockheed following the best practices from the U-2 project that Johnson and Bissell tacitly referred to: complete trust between customer and contractor, individual responsibility and accountability, start-to-finish ownership of design, willingness to take risks, tolerance for failure, and streamlined bureaucracy with minimal staffing and paperwork.

On 29 August 1959, the selection panel voted for the A-12 but required Lockheed to demonstrate by 1 January 1960 that it could reduce the aircraft's RCS sufficiently. CIA awarded a four-month contract to Lockheed to proceed with antiradar studies, aerodynamic structural tests, and engineering designs. Project GUSTO was terminated, and "by a sort of inspired perversity," an Agency officer later wrote, OXCART was selected from a random list of codenames to designate this R&D and all later work on the A-12.(McIninch, 1). The aircraft itself came to be called that as well. Funding for the four-month period was $4.5 million.

During tests over the trial period conducted along with contractor EG&G, Lockheed showed that its concept of shape, nonmetallic parts, and fuel additive would produce the needed reduction in RCS. In the course of this phase of radar testing and after, which required a full-scale, pylon-mounted mock-up, and further wind tunnel tests, the A-12 took on more of its distinctive cobra-like shape that allowed for better dispersion of radar pulses. To further reduce those reflections, the two canted rudders were fabricated from laminated nonmetallic materials, the first time such substances had been used for an important part of an aircraft's structure. (Later on, the production aircraft would be painted with a radar-absorbent coating of ferrite particles in a plastic binder.)

To Bissell's great distress, however, the changes also reduced the aircraft's performance below what he had told the president it could achieve. Johnson had to reduce the A-12's weight by 1,000 pounds and increase its fuel load by 2,000 pounds so it could reach the target altitude of 91,000 feet. He noted in his project log: "We have no performance margins left; so this project, instead of being 10 times as hard as anything we have done, is 12 times as hard. This matches the design number, and is obviously right." (Johnson, Archangel Log, 20).

These modifications worked. On 26 January 1960, Bissell notified Johnson that CIA was authorizing the design, construction, and testing of the new aircraft. Four days later the official word came, and the contract for 12 A-12s was signed on 11 February. Lockheed's original price quotation was $96.6 million, but technical difficulties soon made that figure impossible to meet. CIA included a clause providing for periodic reevaluation of costs. That provision had to be invoked a number of times over the next five years as the A-12's price rose rapidly. With U-2 overflights of the Soviet Union no longer possible after Francis Gary Powers's aircraft was shot down on 1 May 1960, the reliability of the new CORONA satellite still undetermined, and no other aerial or space vehicle considered feasible for the mission, US leaders were willing to pay handsomely to collect vital intelligence on America's principal Cold War adversary.

Lockheed's Aviation Genius

Clarence Leonard "Kelly" Johnson was a pathbreaking aeronautical engineer who worked for Lockheed Aircraft for over four decades. Born in Ishpeming, Michigan, on 27 February 1910, he graduated from the University of Michigan with an M.S. in aeronautical engineering in 1933 and joined Lockheed that same year.

Clarence Leonard "Kelly" Johnson (CIA photo)

Johnson was creative, dynamic, ambitious, and unafraid to question others' expertise and ideas. Soon after arriving at Lockheed, he told his employers that the design of a new aircraft they were working on was flawed and would make the plane dangerously unstable. Instead of firing him, Lockheed asked him to work on the problem. He developed the double vertical tail configuration that became one of the trademark features of the company's aircraft, including the A-12.

From there, Johnson rose quickly and became Lockheed's chief research engineer in 1938. In 1952 he was appointed chief engineer of the firm's Burbank, California, plant and then vice president of research and development in 1956, and vice president for advanced development projects in 1958. Johnson became a member of Lockheed's Board of Directors in 1964 and senior vice president of the corporation in 1969. He retired in 1975, but he served on as a consultant at the Skunk Works until 1980. He died at the age of 80 on 21 December 1990.

Johnson's contributions to advanced aircraft design were extraordinary. He designed or contributed significantly to the development of 40 well-known and important military and civilian aircraft. In addition to the U-2, the A-12, and the SR-71, they included the P-38 Lightning, the Constellation, the PV-2 Neptune, the F-80 Shooting Star, the F-94 Starfire, the F-104 Starfighter, the B-37 Ventura, the C-130 Hercules, the C-140 Jetstar, and the AH-56 Cheyenne attack helicopter. His accomplishments were founded on a hard-charging but informal management style and an openness to experimentation that brought out the best in his coworkers. Among his numerous awards and honors from industry, professional societies, and the Air Force, are two Collier Trophies (1959 and 1964), the National Medal of Science (1966), the nation's highest civilian award, the Presidential Medal of Freedom (1967), the CIA Distinguished Intelligence Medal (1975), and the National Security Medal (1983). He was elected to the Aviation Hall of Fame in 1974.

Earliest Design Concepts

ARCHANGEL 1
JULY 1958

Length:	116.67 ft	Zero Fuel Weight: 41,000 lbs	Cruise Mach:	3.0
Span:	49.6 ft	Fuel Weight: 61,000 lbs	Cruise Alt: 83 - 93 kft	
Height:	23.58 ft	Takeoff Gross: 102,000 lbs	Radius: 2,000 NM	

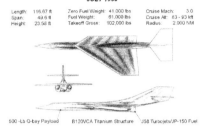

500 -Lb Q-bay Payload   B120VCA Titanium Structure   J58 Turbojets/JP-150 Fuel

ARCHANGEL 2
SEPTEMBER 1958

Length:	129.17 ft	Zero Fuel Weight: 54,000 lbs	Cruise Mach:	3.2
Span:	76.68 ft	Fuel Weight: 81,000 lbs	Cruise Alt: 94 -105 kft	
Height:	27.92 ft	Takeoff Gross: 135,000 lbs	Radius: 2,000 NM	

75" Dia Ramjets Burning HEF
(Lit @ Mach 0.95, 36,000 ft)

Reduced Wing Sweep
Compared to A-1

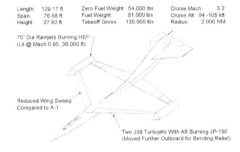

Two J58 Turbojets With AB Burning JP-150
(Moved Further Outboard for Bending Relief)

CONVAIR FISH
"FIRST INVISIBLE SUPER HUSTLER"
1958-1959

Cruise Mach Number:	4.2
Cruise Altitude:	90,000 ft
Range:	3,900 NM
Span:	37.0 ft
Length:	48.5 ft
Height:	9.8 ft

Propulsion:
• Two "Pop-out" Turbojets for Landing
• Two Ramjets for Cruise

Launched From B-58 at Mach 2.2
Above 35,000 Ft

Intermediate Design Concepts

A-5
DECEMBER 1958

Length:	46.0 ft	Zero Fuel Weight:	18,500 lbs	Cruise Mach: 3.2
Span:	32.5 ft	Fuel Weight:	31,820 lbs	Cruise Alt: 90 kft
Height:	16.92 ft	Takeoff Gross:	50,320 lbs	Radius: 1,557 NM

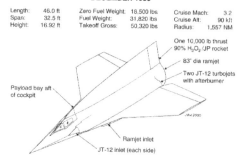

One 10,000 lb thrust
90% H₂O₂ /JP rocket

83" dia ramjet

Two JT-12 turbojets
with afterburner

Payload bay aft
of cockpit

Ramjet inlet

JT-12 inlet (each side)

A-10
FEBRUARY 1959

Length:	109.5 ft	Zero Fuel Weight:	33,300 lbs	Cruise Mach: 3.2
Span:	46.0 ft	Fuel Weight:	52,700 lbs	Cruise Alt: 90.5 kft
Height:	19.25 ft	Takeoff Gross:	86,000 lbs	Radius: 2,000 NM

Significant Improvement Over A-1:
- 18,000 Lb Reduction in TOGW
- 2,500 Ft Additional Altitude

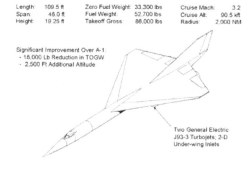

Two General Electric
J93-3 Turbojets; 2-D
Under-wing Inlets

A-11
MARCH 1959

Length:	116.67 ft	Zero Fuel Weight:	36,800 lbs	Cruise Mach: 3.2
Span:	56.67 ft	Fuel Weight:	55,330 lbs	Cruise Alt: 93.5 kft
Height:	21.03 ft	Takeoff Gross:	92,130 lbs	Radius: 2,000 NM

Originally Designed to Carry
31,000 Lbs of HEF and 17,000
Lbs of JP-150

Key Operational Concept: Single
Base + Air Refueling (13,340 NM
Range With 2 Air Refuelings/8
Hour Total Mission Time)

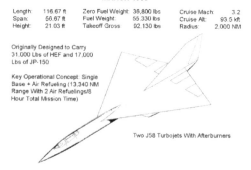

Two J58 Turbojets With Afterburners

The Finalists

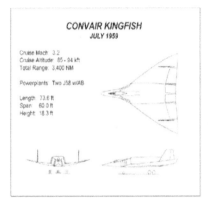

CONVAIR KINGFISH
JULY 1959

Cruise Mach: 3.2
Cruise Altitude: 85 - 94 kft
Total Range: 3,400 NM

Powerplants: Two J58 w/AB

Length: 73.6 ft
Span: 60.0 ft
Height: 18.3 ft

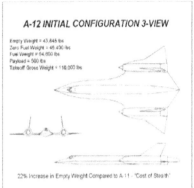

A-12 INITIAL CONFIGURATION 3-VIEW

Empty Weight = 43,846 lbs
Zero Fuel Weight = 45,400 lbs
Fuel Weight = 64,600 lbs
Payload = 500 lbs
Takeoff Gross Weight = 110,000 lbs

22% Increase in Empty Weight Compared to A-11 - "Cost of Stealth"

Finding a Test Site

Lockheed could not do R&D work on the A-12 at its Burbank plant because the runway there was too short and the facility was too public. Ideally, the test site would be remote from metropolitan areas and aviation flight routes, easily accessible by air, fairly close to an Air Force installation, able to accommodate large numbers of personnel, and have good weather all year, fuel storage facilities, and a runway at least 8,000 feet long. No such place existed. After considering 10 Air Force bases slated for closing, as well as Edwards Air Force Base in California, Bissell picked a secluded facility in Nevada. It lacked adequate personnel accommodations and fuel storage, and its runway was too short, but security conditions were optimal, and the needed structures could be built fairly easily.

To test its radar cross section, a full-size model of the A-12 was placed in various positions on a pylon as radar readings were taken. (CIA Photo.)

Construction began in September 1960, with a weekly air shuttle ferrying work crews from Burbank to Las Vegas and then to the test site. Daily commuting was not feasible. Workers at first lived in trailers in Spartan conditions. Later on, surplus Navy housing structures were transported to the base and assembled, and more amenities were added for the residents. By November the new runway had been completed, the old one could not handle the A-12's weight, but the fuel tank farm was not ready until early 1962. Until then, the 500,000 gallons of aircraft fuel needed each month were trucked in. All essential facilities at the site were ready for the delivery of the first A-12 scheduled for August 1961. By the time the testing ended more than four years later, the site had a population of over 1,800, and three shuttle flights arrived every day from Burbank and Las Vegas.

CIA's A-12 "drivers" and managers: (l. to r.) Layton, Sullivan, Vojvodich, Barrett, Weeks, Collins, Ray, BGen Ledford, Skliar, Perkins, Holbury, Kelly, and squadron commander Col. Slater.(CIA Photo.)

Picking the OXCART Drivers

The pilots who would fly the A-12 also had to satisfy a rigorous set of "design specifications." The Air Force, with advice from Kelly Johnson and CIA, drew up the selection criteria. Pilots had to be currently qualified and proficient, with at least 2,000 total flight hours, 1,000 of them in the latest high-performance fighter jets; married, emotionally stable, and well motivated; between 25 and 40 years old; and under six feet tall and 175 pounds so they could fit in the A-12's cramped cockpit. Extensive physical and psychological screening of Air Force personnel files produced 16 candidates. CIA put this group through an intensive security and medical review. The process was kept so secret that the candidates' superiors did not know what their subordinates were doing. Those who survived the screening were approached to work for the Agency on a highly classified project involving a very advanced aircraft. By November 1961, only five had agreed.

A second search and screening produced six more. The 11 pilots selected to fly missions in the A-12 were Kenneth B. Collins, Ronald J. "Jack" Layton, Francis J. "Frank" Murray, Walter L. Ray, Russell J. Scott, William L. Skliar,

Dennis B. Sullivan, Mele Vojvodich Jr., Alonzo J. "Lon" Walter, Jack W. Weeks, and David P. Young. These pilots, known as "drivers," like their U-2 counterparts—were then "sheepdipped" from Air Force to CIA employment, with compensation and insurance arrangements similar to those provided for the U-2 pilots. Scott, Walter, and Young left the program before the A-12 became operational. Skliar was attached to the Air Force portion of OXCART that developed a supersonic fighter-interceptor, the YF-12A, and did not fly the A-12 operationally. In addition, Ray W. Schrecengost flew the two-seat trainer version of the A-12. Lockheed test pilots were: Jim Eastham, Bob Gilliland, Darrell Greenamyer, Bill Park, Art Peterson, Lou Schalk, and Bill Weaver. After the A-12 was decommissioned in 1968, the six surviving operational pilots, Collins, Layton, Murray, Skliar, Sullivan, and Vojvodich, returned to the Air Force. (Ray and Weeks died in A-12 crashes in 1967 and 1968.) Collins and Skliar flew that service's version of the A-12, the SR-71 Blackbird, as test or instructor pilots, and Layton flew the YF-12A.

Patches made for OXCART crew members. CYGNUS was the name given the A-12 in testing. The 1129th SAS was the unit designation of the A-12 team assigned to Kadena Air Base in Okinawa.(CIA Photo.)

Partnership with the Air Force

In addition to providing pilots and assisting with their processing, the Air Force filled several indispensable roles in supporting the OXCART program. The service dispatched to the test site more than a dozen aircraft that were used for training and proficiency flights, cargo transport, search and rescue, administrative travel, and chase flights. (Two of the chase planes and their pilots were lost during the program's testing phase.) The A-12 consumed huge amounts of fuel, 22,000 pounds per hour at cruising speed and altitude, and had to be refueled during its missions. Massive amounts of fuel had to be positioned at special tank farms at several air bases outside the contiguous United States: Eielson, Alaska; Thule, Greenland; Kadena, Okinawa; and Adana, Turkey. The Air Force's 903rd Air Refueling Squadron at Beale Air Force Base was given KC-135 tankers for the refueling operations, and the Air Force detailed most of the support personnel and facilities at Kadena Air Base for Operation BLACK SHIELD, the reconnaissance activity the A-12 would undertake in East Asia. Also, the North American Air Defense Command established procedures so that its radar stations would not report detections of high-performance aircraft. Lastly, Air Force orders for the YF-12A and SR-71 variants of the A-12 helped lower development and procurement costs on the OXCART program.

BREAKING THROUGH TECHNOLOGICAL BARRIERS

"The A-12 'practically spawned its own industrial base' and the 2,400 or so machinists, mechanics, and fabricators could do their own milling and forging." The sign at the top, "Stamp out F.O.D." was an exhortation to "Stamp Out Foreign-object Damage," a problem of engine failure sometimes caused by small objects inadvertently dropped and left in nacelles during fabrication.

Completion of the first A-12 was delayed several times because the performance specifications it had to meet put Johnson and the Skunk Works in uncharted territory. The aircraft, over 101 feet long and weighing up to 62 tons fully loaded, had to fly at Mach 3.2, or 2,150 miles per hour, as fast as a rifle bullet, at a mid-range altitude of 91,000 feet. The A-12 was expected to be over four times faster than the U-2 and go almost three miles higher. Moreover, the aircraft had to have the smallest feasible RCS to minimize the risk of detection and shootdown. To meet all these demands would require developing new structural materials, tools, and fabrication techniques, and special fuels, fluids, lubricants, scalants, paints, plastics, electronics, cables, windshields, fittings, fixtures, and tires. Bissell recalled that the A-12 "practically spawned its own industrial base," and the 2,400 or so machinists, mechanics, and fabricators working on the project could do their own milling and forging.(Bissell, 133) No assembly line techniques could be employed; every aircraft was essentially handmade. At the peak of production in the mid-1960s, nearly 8,000 workers were delivering an A-12 or a variant each month.

Getting to that point was long, hard, and expensive. As a consequence of the technical difficulties, the delivery date of the first aircraft began sliding and costs started rising. Originally promised for May 1961, delivery was moved to August and the first flight was moved to December. A vexed Bissell, then preoccupied with the fallout from the Bay of Pigs debacle that he had overseen as head of CIA's covert operations, wrote to Johnson that "[t]his news is extremely shocking on top of our previous slippage.... I trust this is the last of such disappointments short of a severe earthquake in Burbank."(McIninch, 10) But it was not. In July, Johnson wrote in his log that Lockheed was "having a horrible time building the first airplane ...everyone on edge...and we still have a long, long way to go."(Johnson, Archangel Log, 49) To reduce expenditures, already at $136 million by October 1961 and still climbing, project officials decided to reduce production to 10 aircraft, at a total cost of over $161 million, and assigned a top-level CIA aeronautical engineer to work at Lockheed to monitor the program.

Finding the Right Metal

The most formidable set of challenges Lockheed faced was dealing with the great heat produced by air friction at the high speed the A-12 would reach. Most of the aircraft's skin would be subjected to temperatures between 500 and

600 degrees F., and over 1,000 degrees F. at some spots near the engines. No metal used in aircraft production up to then could withstand such heat. The metals that could stand up under the conditions were too heavy for this project.

After evaluating many materials, Lockheed chose to build over 90 percent of the A-12's airframe out of a titanium alloy. It was almost as strong as stainless steel but weighed half as much and could handle the intense heat. (The rest was made of radar-absorbing composite materials.) Lockheed's supplier of the alloy had trouble delivering material of the requisite quality at first, 95 percent was rejected, and the base metal was scarce enough that some had to be obtained covertly from the country with the largest known reserves, the Soviet Union.(Rich and Janos, 203)

Titanium proved to be very difficult to work with, however. Its extreme hardness caused problems in machining and shaping the material. Drills broke and tools snapped, and new ones had to be devised. By the end of the program, drill bits could make 100 holes before resharpening. Titanium also was very sensitive to contaminants such as chlorine and cadmium. Pentel pens could not be used to draw on sheets of the metal because their chlorine-based ink left etch marks. Wing panels that were spot welded in the summer failed within six or seven weeks, but those made in the winter lasted indefinitely. The problem was traced to Burbank's water, which was heavily chlorinated in the summer to prevent algae growth but not in the winter. Switching to distilled water to wash the panels after acid treatment prevented a recurrence. When bolt heads dropped off under high heat, Skunk Works troubleshooters found that cadmium-plated wrenches left enough residue to weaken the fittings. Hundreds of tool boxes had to be inspected to get rid of the now-useless implements.

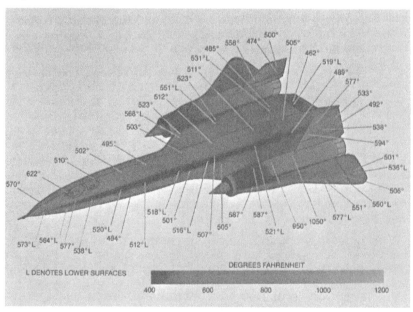

Dealing with the extreme temperatures of Mach 3+ flight was the most formidable challenge.

Fuels, Lubricants, and Sealants

To operate at design speeds and temperatures, the A-12 required fuel, lubricants, hydraulic fluids, and sealants that had not been invented yet. The fuel tanks, holding almost 11,000 gallons, made up the largest proportion of the aircraft and would heat up to about 350 degrees F. At that temperature the most advanced fuel blends then in use would boil off or blow up. Instead, a special fuel, called JP-7, with a low vapor pressure and high flash point had to be developed; a lighted match would not ignite it. A liquid chemical that exploded on contact with air would start the engines. Through the use of heat exchangers and smart valves, the fuel would also act as an internal coolant.

Synthetic lubricants were formulated to work at the extreme temperature range between Mach 3.1 missions and subsonic refueling. They were practically solid at room temperature and had to be heated before each flight. A hydraulic fluid that would not vaporize at high speed but was still usable at low altitudes was eventually found.

No sealant for the fuel tanks was ever developed that was simultaneously impervious to chemical effects caused by the fuel, and elastic enough to expand and contract as the tanks heated and cooled and were subjected to large pressure changes. Consequently the A-12's tanks leaked, a quirk that was not detected until the first aircraft was delivered to the test site and filled with fuel, setting off a reaction that broke down the sealants. A "leak rate" of between five and 60 drops per minute, depending on the source, was considered acceptable. When the A-12 was about to go off on a test or mission, it would receive only enough fuel to get airborne. It would then rendezvous with a KC-135, top off its tanks, and immediately climb to operating altitude, which caused the metal to expand and the leaks to stop.

The Engines

The J58 turbojet engines that would enable the A-12 to fly so high and fast were the most persistent problem. Designed in 1956 for a Navy aviation project that was canceled, the engines had to undergo major modifications to turn them into the most powerful air-breathing propulsion devices ever made. Just one J58 had to produce as much power as all four of the *Queen Mary's* huge turbines, 160,000 horsepower or over 32,000 pounds of thrust. To crank it up, two Buick (later, Chevrolet) racecar engines on a special cart were used. The unmuffled, big block engines put out over 600 horsepower and made a deafening roar. The J58s themselves put out an almost incredible din. Recalling his visit to the test site to watch a midnight takeoff, DCI Richard Helms wrote that "[t]he blast of flame that sent the black, insect-shaped projectile hurtling across the tarmac made me duck instinctively. It was if the Devil himself were blasting his way straight from Hell." (Helms, 266).

As with so much else on the A-12, getting the engines to work at design specifications posed never-before-encountered troubles with fabrication, materials technology, and testing. Not the least of them was the superhot conditions. Maximum fuel temperatures reached 700 degrees F.; engine inlet temperatures climbed to over 800; lubricants ranged from 700 to 1,000; and turbine inlets reach 2,000 degrees F. and above. A Pratt & Whitney engineer later wrote that "I do not know of a single part, down to the last cotter key, that could be made from the same materials as used on previous engines." (Brown, 17)

Pratt & Whitney's continuing difficulties with the weight, performance, and delivery of the J58 forced delays in the completion of the first A-12. After meeting with the manufacturer in early January 1962, Johnson noted in his log that:

> [t]heir troubles are desperate. It is almost unbelievable that they could have gotten this far with the engine without uncovering basic problems which have been normal in every jet engine I have ever worked with... Prospect of an early flight engine is dismal, and I feel our program is greatly jeopardized.(Johnson, Archangel Log, 56)

To prevent further scheduling setbacks, Johnson and CIA officials already had decided to use the less powerful J75 in early flights. The airframe had to be slightly altered to accommodate the substitute engine, which could power the craft only up to Mach 1.6 and 50,000 feet. Despite enormous development costs of the J58, the engines were not ready until January 1963, and the A-12 did not reach Mach 3 speed until the following July, more than a year after the first test flight.

The design feature that ultimately made it possible for the J58s to generate the power needed to fly at planned speed was a pair of retractable, spike-shaped cones that protruded from the engine inlets. The "spikes," as they were known, served as regulators that would decelerate, compress, and superheat incoming air, which was further squeezed and heated by the bypass engines before fuel was added. The supercharged air was then expanded through the turbine and fed into the afterburners. This gas-air mix combusted at 3,400 degrees F., just 200 degrees below the maximum temperature for burning hydrocarbon fuels. Without the spikes, the J-58s would have produced only about 20 percent of the power the A-12 needed. Ben Rich, Lockheed's lead propulsion engineer, recalled that "developing this air-inlet control system was the most exhausting, difficult, and nerve-racking work of my professional life."(Rich and Janos, 209) Rich and his colleagues did much of the testing in wind tunnels at a NASA facility in northern California. They had to work mostly at night because the tests drained too much electricity from the local power grid during the day.

The main issue with the inlets was that the system's mechanical controls did not respond quickly enough to shock-wave-induced variations in the incoming air flow to prevent engine "unstarts" that would cause violent buffeting and severe yawing, and sometimes smash the pilot's head against the cockpit. The unstarts and the "popped shocks" occurred at speeds between Mach 2.5 and 2.9 while the aircraft was on an accelerated climb to design speed. After more than a year and a change in subcontractors, a new electronic control was developed that, along with some other modifications, took care of the problem at lower speeds, but ultimately the inlet system had to be redesigned. In the new configuration, the spike could be moved in or out as much as 26 inches at supersonic speeds to capture and contain the shock wave.

In this photo, one of the first of the A-12 released by CIA, the adjustable inlet cones in front of the engines are clearly visible. Called "spikes," the devices regulated the incoming air flow to maximize thrust and prevent interruptions in fuel combustion at high speeds.

The Cockpit and Flight Suit

Providing for the pilots' safety and comfort was difficult because the external temperatures would make the uninsulated cockpit feel like the inside of a moderately hot oven. To cut weight, Lockheed did not even try to insulate the aircraft's interior; instead, it counted on the pilot's suit to protect him. Pilots would have to wear a type of space suit with its own cooling, pressure control, oxygen supply, and other life support capabilities. Two Lockheed subcontractors, the David Clark Company and the Firewel Corporation, developed a full-pressure suit and oxygen supply system based on ones created for pilots of the X-15 rocket aircraft. The aluminized suit and breathing apparatus would protect the pilot from heat radiated from the 400 degree F. windshield and the effects of depressurization and extreme cold encountered during a high altitude bail-out. The S-901 suits were custom-made and each cost $30,000 in the mid-1960s.

To further protect the pilots, the cockpit had an air conditioning system. It was tested by putting a pilot inside what one engineer described as "a broiler big enough to roast an ox medium rare"(Ibid., 216), a cylinder was cooled to 75 degrees F. while the outer skin was heated to about 600 degrees F. Additionally, if the pilot had to eject from the cockpit, his feet would be held against the seat with cables while it cleared the aircraft, and a stabilization parachute would keep him from spinning and rotating as he descending more than 12 miles in around seven minutes to approximately 15,000 feet, when the main parachute would deploy and separate the pilot from the seat.

The Photo Gear

Notwithstanding its innovations in aeronautical engineering, the A-12 was a photographic reconnaissance platform, so the whole OXCART program would have been pointless if worthwhile pictures could not be taken. Project managers decided to have three different camera systems developed to provide a range of photography,from high-ground-resolution stereo to very-high resolution spotting data.

Perkin-Elmer was the primary manufacturer. Its stereo camera, called Type I, had a 5,000 foot film supply and produced pairs of photographs covering a 71-mile swath with a ground resolution of 12 inches. To meet severe design constraints on size, weight, thermal resistance, coverage, and resolution, Perkin-Elmer employed concepts never before used in camera systems. Perkin-Elmer's camera was installed on all 29 A-12 missions and failed only once, halfway through a sortie.

In case Perkin-Elmer ran into production problems, Eastman Kodak was also asked to build a camera. Called Type II, it had an 8,400-foot film supply and produced stereo photographs covering a 60-mile swath with 17-inch resolution. A third firm, Hycon, built an advanced version of the spotting camera used on the U-2. Hycon's device, Type IV, had a 12,000-foot film supply and covered a 41-mile-wide swath with a resolution of eight inches.

The integrity of the double quartz camera window demanded special attention because optical distortion caused by the effect of great heat (550 degrees F.) on the outside of the window and a much lower temperature (150 degrees F.) on the inside could keep the cameras from taking usable photographs. Three years and $2 million later, the Corning Glass Works came up with a solution: the window was fused to its metal frame by a novel process using high frequency sound waves.

In addition to the film cameras, other collection devices were developed or planned for the A-12: an infrared camera, a side-looking radar, a gamma spectrometer, and a particulate sampler. None of these was used on an A-12 mission.

Defenses

Finally, although it was intended to fly too high and too fast to be detected or shot down, the A-12 was equipped with several electronic countermeasures (ECMs) to foil hostile air defenses. The ECMs would warn the pilot his aircraft had been "painted" by a radar or missile guidance, and then jam or confuse them.

In the Air: "A Wild Stallion"

At last, the first A-12, known as Article 121, was built and ground tested in Burbank during January and February 1962. Because the aircraft was too secret to fly to the test site and too large to carry on a cargo plane, it had to be trucked. During the night of 26 February, a specially designed trailer truck loaded with a huge crate (35 feet wide and 105 feet long) containing the disassembled aircraft's fuselage left the Skunk Works for the two-day trip to the Nevada facility, escorted by the California and Nevada highway patrols and CIA security officers. The box was so wide that some road signs had to be removed, trees trimmed, and road banks leveled. The wings were shipped separately and attached on site.

The A-12's first flight, unofficial and unannounced in keeping with a Lockheed tradition, took place on 25 April 1962 and almost caused the loss of the only OXCART aircraft built so far. Lockheed test pilot Lou Schalk flew the plane less than two miles, at an altitude of about 20 feet, because serious wobbling, Johnson described the movements as "lateral oscillations which were horrible to see," caused by improper hookup of some navigational controls. Instead of circling around and landing, Schalk put it down in the lake bed beyond the end of the runway. When the A-12's nose appeared out of a cloud of dust and dirt, Johnson's angry voice erupted over the radio, "What in Hell, Lou?" (Johnson, Archangel Log, 64; Rich and Janos, 219)

The next day, Schalk tried again, this time with the landing gear down, just in case. The flight lasted about 40 minutes. The takeoff was perfect, but after the A-12 got to about 300 feet it started shedding all the "pie slice" fillets of titanium on the left side of the aircraft and one fillet on the right. (On later aircraft, those pieces were paired with triangular inserts made of radar-absorbing composite material.) Technicians spent four days finding and reattaching the pieces. Nonetheless, the flight pleased Johnson. "We showed that the first flight troubles were not caused by basic aircraft [in]stability."(Johnson, Archangel Log, 65)

Once the fillets were repaired, Article 121 was rolled out for its first official flight on 30 April, just under one year later than originally planned. A number of senior Air Force officers and CIA executives, including Deputy Director for Research Herbert Scoville and former project chief Bissell (who left the Agency in February 1962), witnessed the long-awaited event. Schalk again was the pilot. He took the aircraft up for 59 minutes and reached 30,000 feet and just under 400 mph; most of the flight was made at under 300 mph. He reported that the A-12 responded well and was extremely stable. Johnson said this was the smoothest official first flight of any aircraft he had designed or tested. On 4 May, with Schalk at the controls, Article 121 made its first supersonic flight, reaching Mach 1.1 at 40,000 feet. Problems were minimal. DCI John McCone, who had shown a keen interest in the OXCART program since becoming director in November 1961, sent Johnson a congratulatory telegram.(Scoville telegram to CIA Office of Special Activities (OSA) Deputy Director James A. Cunningham (Cable 4377, IN 35479), 30 April 1962; Scoville memorandum to McCone, "OXCART Program," 7 May 1962.)

Now began the arduous and often discouraging task of bringing the aircraft, christened "Cygnus" after the swan constellation in the northern sky, up to operational performance requirements. Ben Rich later called the A-12 "a wild stallion of an airplane. Everything about it was daunting and hard to tame...so advanced and so awesome that it easily intimidated anyone who dared to come close." (Rich and Janos, 220)

By the end of 1962, four more aircraft had arrived at the test site, and two were engaged in flight testing. Article 122, which arrived in late June, initially was used principally for checking electronic and propulsion systems and RCS. Its first test flight was in January 1963. During its delivery, a Greyhound bus traveling in the opposite direction grazed the 35-foot-wide crate carrying a portion of it. Project managers quickly authorized the payment of nearly $5,000 for damage to the bus so that no insurance or legal inquiry would take place and compromise the program. Articles 123 and 125, after arriving in August and December, respectively, were outfitted for operational use.

Article 124, a trainer version nicknamed the "Titanium Goose," was delivered in November. It was fitted with the less powerful J75 engines, could only reach Mach 1.6 and 40,000 feet, and was the only A-12 that Kelly Johnson ever flew in. (The CIA's first deputy director for science and technology, Albert "Bud" Wheelon, also took a ride in the trainer to demonstrate his confidence in the A-12. John McCone, the director of central intelligence, "roundly criticized" him for "risking my person" that way, Wheelon recalled.)(Wheelon, 76)

The remaining 10 aircraft in the fleet would arrive at the test site by mid-1964. Of those, eight were designated for reconnaissance missions, and two would become the "mother ships" for the D-21 drones in the TAGBOARD project.

Hauled disassembled and in boxes to its Nevada test site, the A-12 posed a significant traffic hazard. Once, an oncoming bus grazed a crate. (CIA Photo.)

FULL STRESS TESTING

Throughout testing, flight procedures evolved; in effect, pilots were testing and training simultaneously. Here, the only A-12 trainer built, the "Titanium Goose," is about to refuel, a process that took the A-12 close to stall speed when it was filled up.

Most test flights were short, averaging scarcely an hour. Through 1963, 573 flights had taken only 765 hours. More air time was not necessary for the earlier tests, and brief flights helped maintain security. Project and test pilots and systems engineers closely critiqued each flight, constantly reviewed data and procedures, and regularly made changes to the latter, in flight and during debriefings afterward. The pilots in effect were performing flight testing and training themselves at the same time. They kept extremely busy in the tight-fitting cockpit, seated amid hundreds of dials, switches, buttons, gauges, and breakers, keeping control of the aircraft with a three-button stick and adjusting for variations in sensitive performance indicators, while navigating at speeds far faster than they had ever flown before. For such a state-of-the-art aircraft, the instrumentation was surprisingly old-fashioned, in keeping with Johnson's preference for tried-and-true systems.

The pilots also practiced aerial refueling with Air Force tanker crews. The first successful hook-up occurred in early 1963. An Agency engineer on the project recalled that connecting with the boom hanging from the back of the KC-135s at around 30,000 feet was tricky. The tanker had to fly as fast as it could, while the A-12 was throttled way back, practically stalling out when it filled its tanks.(Norman and Nelson, quoted in Rich and Janos, 223.)

Some achievements came quickly. The first supersonic flight occurred within a week after testing began, and a speed of Mach 2.16 and an altitude of 60,000 feet were achieved in November 1962. But further progress could not be made because of delays in delivering the J58 engines and inadequacies in those that arrived. "By the end of the year," complained McCone, "it appears we will have barely enough J58 engines to support the flight test program adequately."(McIninch, 13) One of the two flight test aircraft used two J75 engines, and the other used one J75 and one J58. The first A-12 equipped with two J58s flew on 15 January 1963. Ten of the engines were available by the end of the month, and from then on all A-12s but the trainer were fitted with the required propulsion system.

Other performance benchmarks were reached slowly because of continued difficulty sustaining Mach 3 speeds. The first flight to fly briefly at Mach 3 was in July 1963, and the first sustained flight at operational conditions— Mach 3.2 at 83,000 feet for 10 minutes—did not take place until February 1964. A year later, the A-12 fleet had made

1,234 flights totaling almost 1,745 hours, but only 80 of them had been at Mach 3 or faster (one reached Mach 3.27) and for a total of only just over 13 hours.(Briefing Note for the Director of Central Intelligence, Subject: OXCART Status Report," 26 February 1965.

Progress came more quickly during the rest of 1965 and into 1966. Following a one-hour and-forty-minute, 3,000-mile flight mostly above Mach 3.1 in January 1965, an operationally outfitted A-12 (Article 128, not a test aircraft) first attained Mach 3 in March. Peak speed and altitude, Mach 3.29 (over 2,200 mph) at 90,000 feet, were reached by separate aircraft in May and August; 289 flights above Mach 3 lasting over 84 hours were made by October; a maximum stress flight of nearly six-and-a-half hours was completed in November, with portions at peak speed and altitude; and as of mid-March 1966, over 146 out of nearly 2,750 hours flown were above Mach 3.([OSA,] "Project OXCART and Operation BLACK SHIELD Briefing Notes," 20 October 1965; Memorandum from CIA Acting Director of Special Activities to CIA Assistant to the Director of Reconnaissance, "OXCART Status Report," 18 March 1966; [OSA,] OXCART Development Summary and Progress (1 October 1966–31 December 1966)," 31 December 1966;

OSA, "Report - OXCART A-12 Aircraft Experience Data and Systems Reliability," 15 January 1968)

Not surprisingly, people living around the test site and along the flightpaths filed many complaints about sonic booms, especially after the public announcement about the OXCART project in February 1964. Another consequence of all this flight activity was an increase in UFO reports. As with the U-2 in the 1950s, there is a strong correlation between the A-12 flight schedule and "alien aircraft" sightings submitted in the early and mid-1960s. (Haines, 73)

A rare photo of an airborne A-12 with landing gear visible, here on its second flight ever. (CIA Photo.)

Many other hurdles besides the engines had to be surmounted, turning OXCART into a regular "four-alarm fire" that undermined CIA's "reputation for doing things on the cheap [and] quickly," according to Bud Wheelon. (Richelson, 98) In April 1963, after assessing the capabilities of the Soviet Union's new computer equipped TALL KING radar, CIA directed Lockheed to rebuild the chines to change the A-12's RCS, an expensive and, it turned out, undesirable change.(Johnson, "History of the Oxcart Program," 14) Costs soared as a result of other miscalculations, delays, and difficulties. By late November 1963, McCone reported to President Johnson that $400 million had been spent, and $300 million more would be needed, to produce the 15 OXCART aircraft CIA and the Air Force had ordered.(McCone memorandum for the record, "Meeting with the President, Secretary Rusk, Secretary McNamara, Mr. Bundy and DCI," 29 November 1963.)

Some of the problems encountered were mundane, but serious nonetheless. One was foreign object damage, which by July 1963 had resulted in 18 engine removals and extensive nacelle modifications.(Chief/OSA Aircraft Systems Division memorandum for the record, "Factors Affecting A-12 Flight Test and Mach Number Extension," 21 July 1963.) During the aircraft's assembly at Burbank, small items such as bolts, nuts, screws, pens, and metal shavings would fall into the nacelles. When the engines were started at the test site, these objects were pulled into the power plant and damaged its internal parts. Taking X-rays, shaking the nacelles, installing screens over air inlets, and even having workers wear coveralls without breast pockets largely controlled the problem.

Another issue was debris on the taxiways and runway. Like huge vacuum cleaners, the giant J58 engines would suck up anything loose on the pavement, fasteners, clamps, rocks, chunks of asphalt, as they propelled the A-12 toward takeoff. Site personnel had to sweep and vacuum the runway before each test flight.

Although most of the A-12's systems proved acceptably reliable in the less stressful earlier phases of testing, other difficulties arose as the aircraft was put through longer flights at higher speeds and temperatures. As late as March 1965 the inlet control was still a problem, even after well over 10,000 wind tunnel tests, and several months later electrical problems caused by high temperatures persisted. Failed wiring connectors and components incapacitated the inlet controls, communications equipment, ECM systems, and cockpit instruments. The superhot temperatures, structural flexing, vibrations, and sonic shock were more than the materials could stand. Much of the aircraft had to be rewired, and electrical components required redesign.

Crashes

During the first three years of pre-operational testing, three A-12s crashed, two from mechanical malfunctions, one because of ground crew error. All pilots ejected safely. The first loss, of Article 123, occurred on 24 May 1963 during a low-altitude, subsonic flight to test an inertial navigation system. While flying in heavy clouds above 30,000 feet, CIA pilot Ken Collins saw erroneous and confusing air speed and altitude readings just before the A-12 pitched up, stalled, and went into an inverted spin. Unable to regain control of the aircraft, Collins "punched out" at around 25,000 feet. The A-12 spiraled down and crashed.

After parachuting to earth, Collins made his way back to the test site. Wearing a standard fabric flight suit, he avoided having to make difficult explanations to protect OXCART's cover. According to the story given to the press, the accident involved an F-105. The wreckage was recovered in two days, and witnesses were identified and required to sign secrecy agreements. The A-12 fleet was grounded for a week until the cause was determined. So great was the need to find out what went wrong that Collins willingly took truth serum to help his memory. Finally, the inquiry concluded that ice had plugged a tube used to determine airspeed, causing faulty readings that led to the stall and spin.

The next crash occurred on 9 July 1964 while Article 133 was approaching the runway after a Mach 3 check flight. At about 500 feet and 230 mph the aircraft began a steady leftward roll that Lockheed test pilot Bill Parks could not correct. A component of the roll-and-pitch control had frozen. Although only about 200 feet off the ground when he ejected, Parks escaped injury. No news of the accident filtered out of the test site.

On 28 December 1965, barely a month after the A-12 was declared operationally ready, Article 126 crashed less than 30 seconds after takeoff because an electrician had crossed the wiring to the yaw and pitch gyros, in effect reversing the aircraft's controls and making it unflyable. Like Parks, CIA pilot Mele Vojvodich ejected close to the ground but was not injured. DCI William Raborn ordered an investigation into the possibility of sabotage. Simple negligence was found to be the cause, and Lockheed instituted stringent corrective measures. As with the previous crash, there was no publicity about the incident.

Amid increasing concern that the A-12 would not be ready in time for its planned mission to East Asia (Operation BLACK SHIELD), the senior CIA project officer, John Parangosky, met with Kelly Johnson on 3 August 1965 to discuss the problems. They had a frank discussion, and Johnson decided that he needed to assign more top-level supervisors to OXCART and move to the test site himself full time if the A-12's remaining flaws were to be worked out expeditiously. He wrote in his log that:

> *I uncovered many items of a managerial, materiel and design nature...I had meetings with vendors to improve their operation... Changed supervision and had daily talks with them, going over in detail all problems on the aircraft...Increased the supervision in the electrical group by 500%...We tightened up inspection procedures a great deal and made inspection stick.*
>
> *It appears that the problems are one-third due to bum engineering...The addition of so many systems to the A-12 has greatly*

complicated the problems, but we did solve the overall problem.(Johnson, Archangel Log, 98-99)

HIDING OXCART IN PLAIN SIGHT

YF-12A (U.S. Air Force Photo.)

While the A-12 was being tested and refined, US officials mulled over two major issues concerning it. The first was whether to publicly disclose the OXCART program. The Department of Defense had grown concerned that it could not overtly explain all the money the Air Force was spending on its versions of the A-12. At the same time, some CIA and Pentagon officials recognized that crashes or sightings of test flights could compromise the project. With a turning radius of no less than 86 miles at full speed, the A-12 overflew a vast expanse of unrestricted territory. Soon after the first flights in April 1962, CIA and the Air Force changed the program's cover story from involving an interceptor aircraft to a multipurpose satellite launch system.(Scoville to Joseph

Charyk (Undersecretary of the Air Force), "Interdepartmental Cover Support for Project OXCART," 29 May 1962.)

In late 1962 and early 1963 the Department of Defense considered surfacing the YF-12A to provide a cover, reasoning that divulging the existence of a purely tactical aircraft would not reveal any clandestine collection capabilities. Voiced principally by CIA officials and James Killian and Edwin Land of the President's Foreign Intelligence Advisory Board (PFIAB), the contrary argument, disclosing any version of the A-12 would compromise its design innovations, enable the Soviets to develop countermeasures, and destroy its value for reconnaissance, prevailed for the time being. The surfacing issue lingered, however, because OXCART technology would be useful for the Air Force's supersonic B-70 bomber then under development, and for the proposed commercial supersonic transport that Congress was thinking about subsidizing. President Kennedy told CIA and the Pentagon to develop a plan for surfacing the OXCART program but to wait further instructions before proceeding.

By early 1964 the argument for disclosure had become persuasive. More A-12s were arriving at the test site and making more flights. The aircraft's existence probably would be revealed eventually under circumstances the US government could not control, such as a training accident or equipment malfunction, or through a news leak. Commercial airline crews had sighted the A-12 in flight, and the editor of *Aviation Week* indicated that he knew about highly secret activities at the Skunk Works and would not let another publication scoop him. A key factor was that the Soviets' TALL KING radar would be able to identify and track the A-12 despite its small, nonpersistent radar return. Finally, the White House's reluctance to resume flights over Soviet territory would soon force a change in the A-12's mission. Instead of flying over denied areas to collect strategic intelligence, it would most likely be used as a quick-reaction surveillance platform in fast-moving conflicts, a tactical function the Air Force should carry out, not CIA.(McCone untitled memorandum to DDCI Marshall Carter, 10 February 1964; "Briefing Note for the Deputy Director of Central Intelligence...Factors Influencing Decision to Surface the A-11," 10 March 1964.)

On 29 February 1964, the National Security Council decided to surface OXCART. Later that day, the White House announced the successful development of an advanced experimental aircraft, the A-11, which has been tested in sustained flight at more than 2,000 miles per hour and at altitudes in excess of 70,000 feet. The performance of the A-11 far exceeds that of any other aircraft in the world today. The development of this aircraft has been made possible by major advances in aircraft technology of great significance for both military and commercial applications. The A-11 aircraft now at Edwards Air Force Base are undergoing extensive tests to determine their capabilities as long-range interceptors. (Public Papers of the Presidents of the United States: Lyndon B. Johnson, 1963-64, 1:322-23.)

For security reasons, the Air Force's YF-12A interceptor was surfaced, not the A-12, and it was referred to as the A-11, at Kelly Johnson's suggestion. None of the aircraft were already at Edwards, so two had to be rushed from the test site to support the cover story. Johnson recalled that "the aircraft were so hot that when they were moved into the new hanger the fire extinguishing nozzles came on and gave us a free wash job."(Johnson, "History of the Oxcart Program," 15-16) Testing of the A-12s continued at the secret facility; CIA's involvement in the project remained classified, although it was widely assumed.

Surfacing the "A-11" unexpectedly embroiled program managers and technicians in a debate over using an OXCART aircraft to publicly set a world speed record. The presidential announcement stated that "[t]he world record for aircraft speed, currently held by the Soviets [1,665 mph], has been repeatedly broken in secrecy by the...A-11. The President has instructed the Department of Defense to demonstrate this capability with the procedure which, according to international rules, will permit the result of the test to be entered as a new world record." CIA leaders strongly opposed using any of the A-12s to attempt this aeronautical feat. Of the four aircraft used in test flights, only Article 121 had reached the cited speed. Using it in the record trials would set back the testing schedule, jeopardize the aircraft, and undermine the security of the program because the differences between the CIA and Air Force versions would be noticed, and the record would have to be set under the auspices of an uncleared international aviation organization. (Jack C. Ledford (Assistant Director, OSA) memorandum to Wheelon, "Effect on OXCART Program if Aircraft S/N 121 is Used for Speed Record Attempt," 19 August 1964; Ledford memorandum to McCone, "Effect of Using OXCART 121 for Speed Record Attempt," 20 August 1964; Carter letter to Deputy Secretary of Defense Cyrus R.

Vance, 24 August 1964; Cunningham memorandum to McCone, "Establishment of World Record of Aircraft Speed by the A-11," 28 April 1964.)

Consequently, the A-12 was kept out of the competition. No YF-12As were put forward right away because managers of that program were concentrating on armaments rather than speed. At the time, the interceptor had not flown above Mach 2.6. A plane was not ready for the speed trial for over a year. Then on 1 May 1965, a YF-12A set speed and altitude records of 2,070.1 mph and 80,257.65 feet, the first of many for OXCART aircraft.

FINDING A MISSION

All of the OXCART's operational missions were flown out of Kadena Air Base on Okinawa, shown here. (CIA Photo.)

Facing changed circumstances in relations with the Soviet Union and in US satellite development, US policymakers and intelligence officials had to come to grips with how best to use the A-12 as it neared completion. Its intended purpose, replacing the U-2 in overflights of the Soviet Union, had become less and less likely well before the A-12 was operational. Soviet air defenses had advanced to the point that even an aircraft flying faster than a rifle bullet at the edge of space could be tracked. In any event, President Kennedy had stated publicly that the United States would not resume such missions. DCI McCone was determined to find a use for the aircraft, which he later described as "quite invulnerable except under miraculous circumstances" when it met design specifications.(McCone memorandum for the record, "Discussion at NSC Meeting, 5 May 1964," 5 May 1964.) But he lost the argument then, as well as later, when making the case for deploying the A-12 to help determine whether the Soviets had constructed an antiballistic missile system around Leningrad. By 1965, moreover, the photoreconnaissance satellite programs had progressed to the point that manned flights over the Soviet Union were unnecessary to collect strategic intelligence.

The Cuban missile crisis of October 1962 drew attention to the OXCART program because of the threat the U-2 faced from Cuban air defenses. U-2s regularly overflew the island after nuclear missiles were discovered there in mid-October, but two weeks after the discovery, a U-2 was shot down by a Cuban surface-to-air missile. Regular high-altitude reconnaissance of Cuba might no longer be possible. The A-12 now had a potential mission, and achieving operational status became a priority. Because of continued difficulties in achieving design requirements with the J58 engine, however, the A-12 would have to be flown only at up to Mach 2.8 at below 80,000 feet.

This risky program, codenamed SKYLARK, was accelerated during the summer of 1964, after Soviet Premier Nikita Khrushchev declared that after the US elections in November, U-2s flying over Cuba would be shot down. In August, Acting DCI Marshall Carter ordered that SKYLARK be operationally ready by 5 November 1964 in case Khrushchev carried out his threat. (Carter memorandum to Wheelon, "SKYLARK," 22 August 1964.) A detachment of five pilots and ground crew was organized to validate camera performance and qualify pilots for Mach 2.8 operations. They would have to go into action without the full complement of ECMs, as only one of the several devices planned would be available by the deadline, and Agency technical officers were certain that the Cubans would detect the flights and could shoot down the A-12s.

In the end, Khrushchev's threat was bluster, and the A-12 never was used against Cuba. US officials were still discussing the possibility nearly two years later, however, and

CIA officials regarded Cuban overflights as a potentially productive way to test the A-12's ECMs in a hostile area where weather was a factor. Agency analysts judged that the Soviets most likely would react to the flights privately and in low key. The 303 Committee, the NSC group that reviewed sensitive intelligence operations, rejected the idea because it "would disturb the existing calm prevailing in that area of our foreign affairs."([DS&T,] "Vulnerability of the OXCART Vehicle to the Cuban Air Defense System," 15 September 1964; NRO Acting Director memorandum to Deputy Undersecretary of State for Political Affairs et al., "OXCART Reconnaissance of Cuba," 6 September 1966; Peter Jessup (NSC) memorandum, "Minutes of the Meeting of the 303 Committee, 15 September 1966," 16 September 1966; Wheelon to McCone, "Considerations bearing on OXCART use over Cuba," 7 September 1966; CIA Board of National Estimates to Helms, "Probable Communist Reactions to Use of the OXCART for Reconnaissance over Cuba," 6 September 1966; Pedlow and Welzenbach, 44.)

East Asia was the next area US leaders considered using the A-12. The People's Republic of China (PRC) had successfully tested a nuclear device in October 1964, and US military activity in Vietnam was increasing. Overhead collection would be the most important method for monitoring the Chinese program and the military situation in Vietnam, but satellites did not have a quick reaction capability, and several U-2s and drones had been lost over China. US military and intelligence officials drew up a plan for flying OXCART aircraft out of Kadena Air Base on Okinawa under a program called BLACK SHIELD. The Pentagon made available nearly $4 million to provide support facilities on the island, which were to be ready by early fall 1965.

Meanwhile, North Vietnam was starting to deploy SAMs around Hanoi, and a concerned Secretary of Defense McNamara inquired in June 1965 about substituting A-12s for U-2s for reconnaissance over the North. CIA said that BLACK SHIELD missions could be flown over Vietnam as soon as operational performance requirements were achieved. With an overseas deployment looming, personnel at the A-12 test site went all out to have the aircraft meet mission requirements by late 1965. Improvements came faster than expected.

In August, DCI Raborn, who replaced McCone in April 1965, notified President Johnson that an A-12 had successfully simulated an operational mission with two refuelings and three cruise legs. On each leg the aircraft reached its design cruising speed of Mach 3.1 at altitudes between 80,000 and 90,000 feet. The flight covered a total distance of 7,480 miles in just under five and a half hours; forty percent of that time was spent at cruising speed. Only three minor malfunctions occurred; significantly, none involved the air inlets and electrical systems or were related to high heat.(Raborn, "Memorandum for the President," 20 August 1965.

"The Bird Should Leave Its Nest"

Kelly Johnson's firm managerial hand had gotten OXCART back on track. Four A-12s were selected for BLACK SHIELD, and final validation flights were conducted during the fall. During them, the A-12s flew faster, higher, and longer than ever before. On 12 November 1965, the CIA's director of special activities in charge of the program wrote to the Agency's director of reconnaissance that he was "very pleased to announce that, in my judgment, the A-12 aircraft, its technical intelligence sensors, and its operating detachment are operationally ready.... The detachment is manned, equipped, and highly trained.... The aircraft system is performing up to specifications with satisfying reliability and repeatability." Because of "some as yet unexplainable phenomena at cruise conditions," the A-12 could not fly as far as originally intended, but missions could be designed to take that deficiency into account. By 20 November, the validation flights were complete, and all the pilots were Mach 3 qualified. Two days later, Johnson told the Agency that "the time has come when the bird should leave its nest."(CIA Director of Special Activities to CIA Director of Reconnaissance, "Operational Readiness of the OXCART System," 12 November 1965; McIninch, 23.)

Soon after, CIA's Board of National Estimates (BNE)issued an assessment of the potential political implications of BLACK SHIELD. The Agency's most senior analysts judged that the PRC would quickly track overflights of its territory but would not start a diplomatic controversy about them unless it shot down an aircraft. Doing so would occasion a major political and propaganda campaign, but "[w]e do not believe that OXCART missions, whether or not any aircraft came down inside China, would significantly affect Peiping's broader calculations governing its policy toward the war in Vietnam." North Vietnam, "already subjected to heavy US air attack and reconnaissance...would attach little extra significance to the OXCART operation." Lastly, through various sources, the Soviet Union would soon get a fairly complete picture of the scope of BLACK SHIELD but "would probably take no action and make no representations on the matter."(Board of National Estimates, "Political Problems Involved in Operating OXCART Missions from Okinawa over Communist China and North Vietnam," 29 November 1965.)

Analyses such as the BNE's informed the approval process for proposed OXCART missions. The steps were the same as for U-2 flights: an NSC-level recommendation and a presidential authorization. After the A-12 passed its final tests, in early December the 303 Committee ordered the development and maintenance of a quick-reaction capability by 1 January 1966, with deployment to Okinawa 21 days after the president issued his go order.

Then, nothing happened for more than a year. The 303 Committee approved none of CIA's five deployment requests, submitted with support in most instances from the Joint Chiefs of Staff and the PFIAB. Community analysts continued to believe the Chinese, North Vietnamese, or Soviets would not react publicly and belligerently to the missions. Siding with top State and Defense officials, however, the committee did not believe the intelligence requirements at the time, including warning of Chinese intervention in the Vietnam war, were so urgent as to justify the political risk of basing the detachment at Okinawa or revealing some of the A-12's capabilities to hostile nations.

In addition, some reluctance to use the A-12 was related to the discussion that had already begun about phasing out the CIA program. In mid-August 1966, President Johnson listened to the divergent views and upheld the 303 Committee's decision not to fly actual missions for the time being.(Peter Jessup (NSC) memorandum for the President, "Proposed Deployment and Use of A-12 Aircraft," 11 April 1966; Raborn memorandum to the President, "OXCART Deployment Proposal," 29 April 1966; Raborn memorandum to the 303 Committee, "OXCART Deployment," 15 June 1966; Special National Intelligence Estimate 10-2-66, "Reactions to a Possible US Course of Action," 17 March 1966; "OXCART Development Summary and Progress (1 October 1966-31 December 1966)."

Biding Time, Sharpening Procedures

During these months OXCART personnel worked on refining mission plans and flight tactics, testing the aircraft and systems, training, and preparing the forward base at Kadena. The delay was beneficial. Even though the A-12 had been declared operationally ready, important components in the propulsion system still needed correction. More efficient procedures reduced the time required to go from mission notification to deployment from 21 to 15 days. Six operationally configured aircraft were constantly training and engaging in operational flight simulations.

In October 1966, one week after its first flight, Article 127 flew for seven hours and 40 minutes, the longest time in air so far. Two months later, Lockheed test pilot Bill Parks completed an impressive demonstration of the A-12's capabilities by flying 10,198 miles in six hours at an average speed of 1,659 mph (including slowdowns for refueling), setting a speed and distance record unattainable by any other aircraft. By mid-February 1967, 2,299 test and training flights had been flown over 3,628 hours, with more than 332 of those at Mach 3 or higher.([OSA,] "Briefing Note for the Director of Central Intelligence...OXCART Status Report," 15 February 1967)

The first fatality of the OXCART program occurred on 5 January 1967, when Article 125 crashed, killing CIA pilot Walter Ray. Because of a faulty fuel gauge and related electrical equipment problems, the aircraft ran out of fuel while on its descent to the test site. Ray ejected at between 30,000 and 35,000 feet but did not separate from the seat. That kept the parachutes from deploying, and he fell to earth, dying on impact. To protect the security of the A-12 program, the Air Force informed the media that an SR-71 was missing and presumed down, and identified the pilot as a civilian. Like the three crashes that preceded it, Ray's involved a problem inherent in any new aircraft, a malfunction of a part specifically designed and built for it. None of the four incidents occurred while the A-12 was being subjected to the unprecedented rigors of design speeds and altitudes.

Missions Begin: Spying on the Enemy

By early 1967, the Johnson administration was growing anxious that the North Vietnamese could deploy surface-to-surface missiles (SSM) targeted at the South without being detected. When the president asked for a collection proposal, CIA suggested that the A-12 be used, noting that its camera was better than those on drones or the

U-2, and that it was much less vulnerable than those platforms and more versatile than the CORONA satellites. (Helms memorandum to the 303 Committee, "OXCART Reconnaissance of North Vietnam," with attachment, 15 May 1967.) DCI Helms brought up the idea at a luncheon with the president on 16 May and got his approval. The Agency put the BLACK SHIELD deployment plan into effect later that day.

On 17 May, the airlift of personnel and equipment to Kadena began, and Articles 131, 127, and 129, flown by Vojvodich, Layton, and Weeks, arrived between 22 and 27 May. The first two flew non-stop from Nevada to Kadena; the third diverted at Wake Island to correct an equipment malfunction and finished the trip the next day. The unit, which at its inception had been designated for security purposes as the 1129th US Air Force Special Activities Squadron (SAS), Detachment 1, comprised three A-12s, six pilots (three deployed at a time; later two), and over 250 support personnel. Its commander was Col. Hugh "Slip" Slater, who had worked with CIA on the U-2 program and at the OXCART test site. The 1129th SAS was ready for operations by the 29th. The call came the next day to fly the first mission on the 31st over North Vietnam.

Piloted by Vojvodich (Layton was the secondary, and Weeks was the backup), Article 131 took off just before 1100 local time in a torrential downpour. The A-12 had never operated in heavy rain before, but the weather over the target area was forecast to be satisfactory, so the flight went ahead. It lasted three hours and 39 minutes and was flown at Mach 3.1 at 80,000 feet. Vojvodich crossed the coast of North Vietnam at 1014 local time (Vietnam is two hours ahead of Okinawa), flew the planned single-pass route in less than nine minutes, refueled over Thailand, exited near the Demilitarized Zone (DMZ) at 1122, and touched down at Kadena in the rain at 1233 local time.

The mission was a success, photographing 70 of the 190 known SAM sites and nine other priority targets, including an airfield, a military training area, an army barracks, and the port at Haiphong. No SSM facilities were located. Contrary to some published accounts, neither Chinese nor North Vietnamese radar tracked the aircraft, nor did North Vietnam fire any missiles at it. Those hostile reactions did not occur until the third and 16th missions, respectively.([OSA] "Critique for OXCART Mission BSX001," 6 June 1967; DS&T, "BLACK SHIELD Reconnaissance Missions, 31 May-15 August 1967," 22 September 1967, 3-4; National

Photographic Interpretation Center (NPIC), "BLACK SHIELD Mission X-001, 31 May 1967," NPIC/R-112/67, June 1967; [OSA,] "Critique for OXCART Mission Number BX6705," 26 June 1967, and "Critique for OXCART Mission Number BX6732," 3 November 1967.)

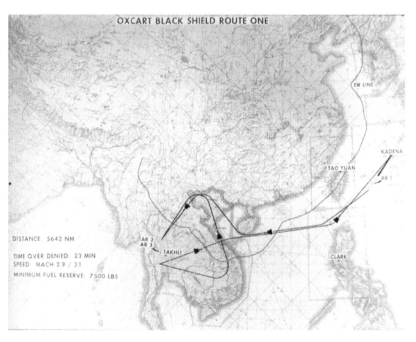

OXCART's first mission over Southeast Asia, 31 May 1967. With pilot Vojvodich in the cockpit, Article 131 refueled three times during its 3 hour 39 minute flight. (CIA Photo.)

This image of Hanoi area was taken during the fourth mission, on 30 June 1967. (CIA Photo.)

Through 6 May 1968, the date of what would become the last flight, the A-12 pilots at Kadena flew 29 missions out of 58 they were put on alert to perform: 24 over North Vietnam; two over Cambodia, Laos, and the DMZ; and three over North Korea. The flights were distributed among the pilots: Collins and Layton had six, Vojvodich and Weeks got five, Murray did four, and Sullivan was on three. The aircraft were flown at between Mach 3.1 and 3.2 and a bit above 80,000 feet. At that height, above the jet stream, air

75

turbulence was minimal, and the curvature of the earth showed beneath the black, star-flecked sky.

The A-12s' aeronautical components and photographic systems proved very reliable. Twenty-seven of the sorties were judged successful, based on the quality of photography returned; two were deemed partially successful or unsuccessful due to cloud cover or a camera malfunction. One mission had to be cut short after one pass because of an engine problem. None of the 29 cancelled alerts were the result of mechanical concerns; bad weather caused all but three, which were due to operational decisions. The A-12s were so fast that they typically spent only about 12.5 minutes over North Vietnam on a single-pass mission and 21.5 on a double-pass route.

Project headquarters in the Washington DC area planned and directed all the A-12 missions. Their preparation followed this procedure: Each day at 1600 local time a mission alert briefing took place. If the weather forecast, the key variable in deciding whether to go ahead or cancel the sorties, seemed favorable, Kadena was alerted and given a flight profile about 28 to 30 hours before takeoff. The primary and back-up aircraft and pilots were selected. The A-12s, painted black and bearing no markings other than red tail numbers that were changed every mission, got thorough inspections and servicing, all systems were checked, and the cameras were loaded into the bays. On the evening before the day of the flight, the pilots received a detailed briefing of the route. Twelve hours before takeoff (H minus 12), headquarters again reviewed the weather over the target. If it was still favorable, preflight procedures continued.

On the morning of the flight, the pilots got a final briefing. The aircraft's condition was reported, weather forecasts were reviewed, and changes in the mission profile and relevant intelligence was communicated. At H minus 2, headquarters issued a "go/no-go" decision. At this point the weather forecast also had to be good for the refueling areas. If the mission was still on, the primary pilot received a medical examination, suited up, and squeezed himself into the aircraft. If any malfunctions developed, the back-up would be ready to fly one hour later. This proved necessary only once. On the second mission on 10 June 1967, the primary A-12 lost a fillet panel during refueling and returned to base, and the back-up completed the mission. On most BLACK SHIELD flights, the A-12s were airborne about four hours. The shortest complete mission in Southeast Asia lasted just over 3.5 hours; the longest took nearly 5.5 hours. The aircraft took on fuel two or three times, depending on the planned route, on each operational flight: once, soon after takeoff south of Okinawa, and once over Thailand for each pass it would make over the target area before it returned to Kadena.

After the A-12s landed, the camera film was removed from the aircraft, boxed, and sent by courier plane to a processing facility. At first the film was developed at the Eastman Kodak plant in Rochester, New York. That trip took too long for US military commanders who wanted the intelligence more quickly. By late summer, processing was shifted to an Air Force center in Japan, and the photography could be available to the US military in Vietnam within 24 hours after a mission was completed.

Less than two months into BLACK SHIELD, analysts had enough evidence to conclude that North Vietnam had never deployed SSMs. By the end of 1967, the A-12 had collected clear, interpretable photography of all of North Vietnam except for a small area along the border with the PRC. The BLACK SHIELD missions provided valuable imagery of SAM sites, airfields and naval bases, ports, roads, and railroads, industrial facilities, power plants, and supply depots. Military planners and photo interpreters used the information to develop air and air defense order of battle estimates, assess bomb damage, and develop flight routes and target sets for bombing runs, enabling US pilots to accomplish their missions more effectively and in greater safety. Analysis of photography of the DMZ gave insights into North Vietnamese infiltration and supply routes and North Vietnamese and Viet Cong troop deployments.(OSA mission critiques, 16 June 1967-15 May 1968; DS&T, "BLACK SHIELD Reconnaissance Missions, 16 August-31 December 1967," 31 January 1968, and "BLACKSHIELD Reconnaissance Missions, 1 January-31 March 1968.")

President Johnson's national security adviser, Walt Rostow, recounted that the A-12 missions (along with those of the SR-71) "were invaluable to the president." Without them, he "would never have allowed any tactical air operations in the North because he was extremely sensitive...to the possibilities of a bomb accidentally hitting a Chinese or Russian ship while it was unloading in the harbor, and he also was determined to keep civilian damage and casualties to a minimum." Johnson "usually chose the targets personally and insisted on approving each and every raid into the North.... Before signing off on a mission he calculated in his own mind whether the anticipated losses were worth the anticipated gains." The A-12 and SR-71 photographs "were the decisive factors in helping him to make up his mind."(Quoted in Rich and Janos, 244.)

Under Fire over Vietnam

North Vietnam fired SAMs at BLACK SHIELD A-12s three times but caused damage only once. The first attempted shootdown occurred on the 16th mission on 28 October 1967. Flown by Dennis Sullivan, the aircraft was on its second pass, approaching Hanoi from the west, when an SA-2 was launched at it. Photographs taken during the mission show missile smoke above the SAM site and the missile and its contrail, heading down and away from the aircraft. The A-12's ECMs worked well, and the SAM, which was fired too late, was never a threat. (DS&T, "BLACK SHIELD Reconnaissance Missions, 16 August-31 December 1967," 31 January 1968, 18-22; D/OSA memorandum to DDS&T, "Analysis of Surface to Air Missile Engagements for OXCART Missions BX6732 and BX6734," 27 November 1967.

The second incident, two days later, on the 18th mission, was the closest an OXCART aircraft ever came to being shot down. Sullivan again was the pilot. On the first pass between Hanoi and Haiphong, radar tracking detected two SAM sites preparing to launch, but neither did. On the second pass toward Hanoi and Haiphong from the west, at least six missiles were fired from sites around the capital. The A-12 was flying at Mach 3.1 at 84,000 feet. Looking out the rear-view periscope, Sullivan reported seeing six vapor trails go up to about 90,000 feet behind the aircraft, arc over, and begin converging on it. He saw four missiles, one as close as 100 to 200 yards away, and three detonations behind the A-12. Six missile contrails appeared on mission photography.

A post-flight inspection at Kadena found that a piece of metal, probably debris from an exploded missile, had penetrated the lower right wing and lodged near the fuel tank. A BLACK SHIELD officer at Kadena noted that the A-12 pilots were "showing considerable anxiety about overflying this area before we get some answers." Helms ordered that missions be temporarily suspended. None was flown until 8 December. It and the following one two days later photographed the Cambodia-Laos-South Vietnam triborder area and were not sent over the North. (DDS&T Carl Duckett memorandum to DCI Richard Helms, "OXCART Operations on 27, 28, 29 October (local time)"; DS&T, "BLACK

SHIELD Reconnaissance Missions, 16 August-31 December 1967," 31 January 1968, 25-35; D/OSA memorandum to DDS&T, "Analysis of Surface to Air Missile Engagements for OXCART Missions BX6732 and BX6734," 27 November 1967; Cable OPCEN 2898, 30 October 1967; Cable from Kadena, IN 91487, 1 November 1967; Donald Smith (EA/DDCI) untitled memorandum to Duckett, 6 November 1967.

Sorties over North Vietnam resumed on 15 December and continued until 8 March 1968, the next-to-last BLACK SHIELD flight. The first two flights took different paths than the Hanoi-Haiphong route followed by the A-12s that were shot at in late October. Another SAM was fired on mission 23 on 4 January 1968; that aircraft took the same route as those that had been attacked. The missile, fired on the second pass like the others, was captured on photography from launch to detonation, well over a mile from the aircraft. Two of the next three flights over North Vietnam came in from the south rather than the east, and all three stayed farther away from Hanoi and Haiphong than those that had been shot at. The general times when these flights were made did not change despite the SAM attacks; all crossed into North Vietnamese territory in the late morning. (DS&T, "BLACK SHIELD

Reconnaissance Missions, 16 August-31 December 1967," 36-39; DS&T, "BLACK SHIELD Reconnaissance Missions, 1 January-31 March 1968," 30 April 1968, 3-8, 10-11, 13-14.)

Looking for the *Pueblo*

The North Korean seizure of the US Navy ship *Pueblo* while it was on a SIGINT mission in international waters on 23 January 1968 enabled the A-12 to demonstrate its superiority as a quick-reaction collection platform. Although the US military had indicated its interest in BLACK SHIELD overflights of North Korea even before the incident, the Department of State had opposed them, and none was planned when the *Pueblo* was captured. Walt Rostow remembered that "[t]he whole country was up in arms over this incident. The president was considering using airpower to hit them [the North Koreans] hard and try to shake our crewmen loose. But when we cooled down, we had to suck in our gut and hold back until we were sure about the situation."

Helms urged use of the A-12 to find the missing ship. Johnson was reluctant at first to offer such a "tempting target" but was assured that the aircraft "could photograph the whole of North Korea, from the DMZ to the Yalu River, in less than

10 minutes, and probably do so unobserved by air defense. (Quoted in Rich and Janos, 245.)

This photo above was taken by Article 131 on 26 January 1968, the only time an A-12 camera took an image of the Pueblo. (CIA Photo.)

On 24 January the *Pueblo* advisory group, comprising senior officials from the White House, the Departments of Defense and State, and CIA, had Helms draw up a reconnaissance plan that included A-12s. President Johnson approved their use later that day.(Summary Minutes of Pueblo Group" and "Notes of the President's Meeting," both 24

January 1968, Foreign Relations of the United States, XXIX, Part 1, Korea, 474, 475, 494; Helms memorandum to Walt W. Rostow et al., "OXCART Reconnaissance of North Korea," 24 January 1968.) On the 26th, Jack Weeks flew a three-pass mission over the southern part of North Korea and the Demilitarized Zone. The purpose was to determine whether Pyongyang, which claimed it had caught the United States spying inside its territorial waters, was mobilizing for hostilities. Chinese radar tracked the A-12, but no missiles were fired during the highly successful mission.

Substantial intelligence was acquired on North Korea's armed forces; no signs of a military reaction were detected; and the *Pueblo*, apparently undamaged, was found in a small bay north of Wonsan accompanied by two patrol boats.(NPIC, "North Korea Mission BX 6847, 26 January 1968, Highlights," NPIC/R-17/68, January 1968; DS&T, "BLACK SHIELD Reconnaissance Missions, 1 January-31 March 1968," 30 April 1968, 8-10.) "So we had to abandon any plans to hit them with airpower," according to Rostow. "All that would accomplish would be to kill a lot of people, including our own." But the A-12's photographs "provided proof that our ship and our men were being held. The Koreans couldn't lie about that, and we immediately began negotiations to get them back. After difficult and protracted discussions, North Korea released the surviving crewmembers 11 months later. (Quoted in Rich and Janos, 245.)

The US military wanted a second overflight of North Korea, but the *Pueblo* advisory group decided not to recommend any more right away because the "excellent" photography taken on the 26th, along with other information, was deemed sufficient to answer the crucial questions. By mid-February, however, the need returned. After the Department of State accepted assurances that it was highly unlikely the A-12 would come down in hostile territory if something went wrong, the 303 Committee approved two more missions over the peninsula. They were flown on 19 February and 6 May. On the first sortie, scattered clouds concealed the area where the *Pueblo* had been spotted. (The ship had been moved by then.) The second flight—the last A-12 mission, as it turned out—was piloted by Jack Layton. Like the other missions over North Korea, it found no sign of a military buildup. (Joseph F. Carroll (Director, DIA) memorandum to Chairman, Joint Chiefs of Staff, "Requirement for a Second BLACK SHIELD Mission Over North Korea," 29 January 1968; "Report on Meeting of the [Pueblo] Advisory Group" and "Notes of the President's

Luncheon Meeting with Senior American Advisors," both 29 January 1968, FRUS, XXIX, Part 1, 557, 565; DS&T, "BLACK SHIELD Reconnaissance Missions, 1 January-31 March 1968," 30 April 1968, 11, and "BLACK SHIELD Reconnaissance Missions, 1 April-9 June 1968," 7 August 1968, 2-3; CIA Intelligence Information Cable, "Implications of Reported Relocation of USS Pueblo," 12 February 1968, Declassified Documents Reference System, doc. no. CK3100137943.)

A FUTILE FIGHT FOR SURVIVAL

With the OXCART program cancelled, the A-12s flew back to the United States and were placed in storage. (CIA Photo.)

Two major ironies run through the history of the A-12. One is that it was never used for its intended purpose of overflying the Soviet Union to collect strategic intelligence on Moscow's nuclear weapons capabilities, and instead was deployed as a tactical collection platform in a conventional military conflict. The other is that just as the A-12 was about to be declared operationally ready, US policymakers had decided to replace it with the Air Force's OXCART reconnaissance variant, the SR-71. The most advanced aircraft ever built was decommissioned after less than a year in service, not from any shortcomings in its design but because of fiscal pressures and competition between the reconnaissance programs of CIA and the Air Force.

The first step in the A-12's phase-out came in November 1965, when the Bureau of the Budget (BOB) expressed concern about the costs of the two programs. It questioned both the total number of aircraft required for the combined fleets and the need for a separate CIA unit. Among several alternatives, BOB recommended closing down the A-12 program by September 1966 and halting acquisition of more SR-71s. It asked CIA and the Pentagon to explore the options. A senior OXCART manager summed up CIA's position: the BOB proposal would "deny the United States Government a non-military capability to conduct aerial reconnaissance of denied areas in the world in the years ahead." Secretary of Defense McNamara rejected the BOB recommendation, probably because the SR-71 would not be mission ready by that date. (Ledford, "Briefing Note for the Director of Central Intelligence...Bureau of the Budget Recommendations for the OXCART Program," 16 November 1965.)

Nothing more happened on the matter until June 1966, when a study group that BOB proposed was set up to look at ways to reduce expenses in the OXCART program. The group, which included representatives of BOB, CIA, and the Pentagon, identified three alternative courses: continue both fleets at current levels; mothball all A-12s while sharing SR-71s between CIA and the Strategic Air Command (SAC); or, assuming the SR-71 was available by then, end the A-12 program in January 1968 and assign all missions to the Air Force fleet. The group noted that for the next several years, both aircraft would remain uniquely capable of conducting tactical reconnaissance missions during periods of international tension and hostilities but that toward the end of the decade certain satellites and drones could supplant them. (C. William Fischer (BOB), Herbert D. Benington (DOD), and John Parangosky (CIA), "Advanced Reconnaissance Aircraft Study," November 1966.)

On 12 December 1966, more than five months before Article 131 would fly the A-12's first operational mission, four senior US officials met to consider the options. Over DCI Helms's objections, Deputy Defense Secretary Cyrus Vance, BOB Director Charles Schultze, and presidential scientific advisor Donald Hornig decided on the third course. The United States would end the A-12 program.

Hoping in some way to keep a supersonic reconnaissance capability for the Agency, Helms then asked that the SR-71s be shared between CIA and the Air Force. After hearing that the SR-71 most likely would not be ready by the time the A-12s were taken out of service and that its performance characteristics made it more vulnerable than the CIA aircraft, the DCI again urged continuation of the A-12 program. In late December, however, President Johnson chose to close it down by 1 January 1968.(Parangosky memorandum for the record, "12 December 1966 Meeting with Mr. Helms re OXCART/SR-71," 15 December 1966; DDCI Rufus Taylor memorandum to Helms, "Reduction of A-12/SR-71 Fleets," 29 December 1966.)

CIA now had to develop a plan for decommissioning the A-12. Project managers informed Vance on 10 January that four aircraft would be placed in storage in June 1967, two more by December, and the last four by the end of January 1968. (The sequencing changed later on.) On the personnel side, over 2,000 government and contractor employees, including 420 from CIA and the Air Force, would also have to be reassigned or dismissed. In May, Vance directed that the SR-71 take over the contingency responsibility for Cuban overflights as of 1 July and that it handle all Southeast Asia missions by 1 December. In the meantime, the A-12 detachment was to maintain its capability to deploy to East Asia in 15 days and over Cuba in seven.([OSA,] "Briefing Note for the Director of Central Intelligence...OXCART Status Report," 15 February 1967; Helms memorandum to Rusk, "OXCART," 28 July 1967; Vance memorandum for Chairman, Joint Chiefs of Staff et al., "SR-71 Plans," 9 May 1967.)

All this planning took place before the A-12 had flown a single mission. Once those began in May 1967 and produced very useful intelligence, and with the SR-71 not ready yet, high-level US officials, notably the president's national security adviser, the PFIAB, and the president's Scientific Advisory Committee, and some members of Congress started having reservations about stopping the CIA program. Administration officials considered extending the A-12's operational mission beyond the end of the year. Helms pressed the Pentagon for a decision soon because maintenance and readiness would suffer with further delay, but the administration took no immediate action. (Helms letter to

McNamara, 13 September 1967; [National Reconnaissance Office,] "NRP Executive Committee, Minutes of Meeting Held September 29, 1967, Office of Deputy Secretary of Defense," 29 September 1967.)

One way to help decide whether to keep one or both aircraft was to determine which performed better. CIA contended that the A-12 did because it flew higher and faster and had superior cameras. The Air Force countered that the SR-71 was preferable for intelligence purposes because it had three different cameras, for area search, spotting, and mapping, and carried sensors the A-12 did not at the time, infrared detectors, side-looking airborne radar, and ELINT-collection devices needed for its mission of post-nuclear-strike reconnaissance.

To resolve the question, the aircraft competed one-on-one in a flyoff codenamed NICE GIRL. Between 20 October and 3 November 1967, A-12s and SR-71s flew three identical routes along the Mississippi River about one hour apart with their collection systems on. Representatives from CIA, the National Photographic Interpretation Center, the Defense Intelligence Agency, and other military intelligence organizations evaluated the data collected. The results were inconclusive. The A-12's camera worked better, it had a wider swath and higher resolution, but the SR-71 collected types of intelligence the CIA aircraft could not, although not yet of very good quality.(Duckett memorandum to Helms, "OXCART/SR-71 Information for EXCOM Meeting," 19 December 1967.) However, some of its sensors would have to be removed to make room for ECM gear, a salient point now that North Vietnam had shot at two BLACK SHIELD aircraft.

Because of its track record and continued delays with the SR-71, the A-12 won a temporary reprieve in late November 1967 when the Johnson administration decided to keep both fleets temporarily. A month later, the Pentagon announced that five A-12s would be kept operational through 30 June 1968 while the SR-71 was prepared to begin missions over North Vietnam "as rapidly as ECM implementation and other program considerations will permit."

With expenditures of the Vietnam War rising steadily, US policymakers revisited the question. Another study of the subject was completed in the spring of 1968. It came up with new scenarios involving combinations of closures, transfers, and decommissioning, along with the Air Force takeover and status quo options. Helms did not bend and argued for the last, underscoring the importance of a covert reconnaissance capability under civilian management.(Helms memorandum to Paul Nitze (DOD) and Hornig, "Considerations Affecting OXCART Program Phaseout," 18 April 1968. 8OSA memorandum to Lockheed, "A-12 Accident Report, Aircraft 129," 21 June 1968.)

On 16 May 1968, however, the new secretary of defense, Clark Clifford, reaffirmed the original decision to end the A-12 program and mothball the aircraft. The president concurred five days later. The A-12 sortie on 8 May would be its last. Agency personnel at Kadena started packing up and preparing to return home. Project headquarters designated 8 June as the earliest date for phasing out all the aircraft. Those at the test site were placed in storage in a hangar at Palmdale, California, and those at Kadena were readied for flights back.

The second pilot fatality in the program occurred during this drawdown. On 4 June 1968 Jack Weeks was in Article 129 on a checkout flight after an engine change for the trip to the United States. He was last heard from 520 miles east of Manila. No trace of the plane was found, and an investigation turned up no clue about the cause of the crash. Signals received about a half hour into the flight from the onboard BIRDWATCHER monitoring system indicated engine trouble; a catastrophic failure was the most likely explanation. (OSA memorandum to Lockheed, "A12 Accident Report – Aircraft 129," 21 June 1968.)

The other two A-12s left Okinawa on 8 and 19 June. Frank Murray made the final flight of an A-12, in Article 131, on 21 June from the Nevada test facility to the California storage site. The only major components of the aircraft that could be salvaged for its successor were the J58 engines. The Perkin-Elmer Type I's were too big to fit in the SR-71's camera bay.

On 26 June, the A-12 operational pilots, Ken Collins, Jack Layton, Frank Murray, Dennis Sullivan, Mele Vojvodich, and Jack Weeks (posthumously), were awarded the CIA Intelligence Star for valor. Other members of the 1129th SAS, nicknamed the Roadrunners, received awards as well. The unit's commander, Col. Slater, and his deputy, Col. Maynard "Am" Amundson, were given the Air Force Legion of Merit, and the Air Force Outstanding Unit Award was presented to all the detachment's members. The pilots' wives attended the ceremony and learned for the first time what their husbands really had been doing for the past several years.

Kelly Johnson, a guest speaker at the event, gave a moving address in which he lamented the demise of the enterprise that represented his pinnacle accomplishment as Lockheed's most creative aeronautical engineer. Two days earlier, he had written that "[i]t's a bleak end for a program that has been overall as successful as this." (Johnson, "Archangel Log," 108.)

Assessing the A-12's Contribution

The value of the A-12 must be determined by two standards: aviation and intelligence. In the first instance, the OXCART program must be judged a categorical success. It produced what it was intended to: a reconnaissance aircraft that could fly at unprecedented speeds and heights for unequaled ranges, and was essentially invulnerable to enemy attack. OXCART represented a pioneering accomplishment in aeronautical engineering. Well over 40 years after it first flew, the A-12's maximum speed and altitude have not been equaled by a piloted operational jet aircraft. The exceptionally demanding design requirements for speed, altitude, and stealth produced innovations in aerodynamic design, engine performance, cameras, metallurgy, use of nonmetallic materials, ECMs, RCS suppression, and life support systems that were used for years after and helped lay the foundation for future stealth research. Finally, no A-12 or its operational successor, the SR-71, was shot down despite hundreds of attempts while they conducted nearly 3,600 operational missions over nearly a quarter century.

As an intelligence collector, the A-12's record is commendable but less striking, although not due to anything about the aircraft itself. US policymakers decided not to use the A-12 for its original purpose. The technological breakthroughs that made the aircraft possible took longer than expected, and between the time Lockheed conceived the idea of the OXCART and the time the aircraft was operationally ready, the international and technological situation had changed. The U-2 shootdown in May 1960 made overflights of the Soviet Union politically unfeasible, and by the early 1960s spy satellites were collecting the required information on Soviet military developments.

US leaders considered use of the A-12 to collect strategic intelligence on the PRC but chose to rely principally on satellites. In the end, the A-12 contributed little to the Agency's strategic intelligence mission. In addition, the complexities of running A-12 sorties, planning routes, mobilizing several hundred personnel, deploying fuel and tankers, and programming guidance systems, made the aircraft very costly to operate and less useful as a quick-response platform than the U-2, which remained in service despite the presence of its replacement.

When it did fly intelligence missions, however, the A-12 performed superbly. During BLACK SHIELD the A-12 acquired timely and usable photography of North Vietnam's air defense network, key military and economic targets, and war-related activities that enabled US military commanders to plan more effective bombing routes while keeping US pilots farther out of harm's way. Analysis of A-12 photography quickly enabled the US government to determine that North Vietnam had no SSMs, dispelling a growing concern that a serious escalation of the war was imminent. And during the tense time after the *Pueblo* seizure, A-12 missions over North Korea allayed US fears that Pyongyang was preparing for military action in the wake of the incident.

As a tactical intelligence collector, the A-12 had a near-perfect record, but it fell victim to budgetary pressures and interdepartmental differences over how best to use the expensive aerial reconnaissance assets of the United States.

As an Agency history of overhead reconnaissance observed, "[t]he most advanced aircraft of the 20th century had become an anachronism before it was ever used operationally." (Pedlow and Welzenbach, 50) Yet the two men most responsible for bringing the A-12 into existence and making it work, the visionary engineer Kelly Johnson and the realist technocrat Richard Bissell, anticipated this outcome. In his project log in 1967, Johnson wrote:

> *I think back to 1959, before we started this airplane, to discussions with Dick Bissell where we seriously considered the problem of whether there would be one more round of aircraft before the satellites took over. We jointly agreed there would be just one round, and not two. That seems to have been a very accurate evaluation.* (Johnson, "Archangel Log," 105.)

REFERENCES

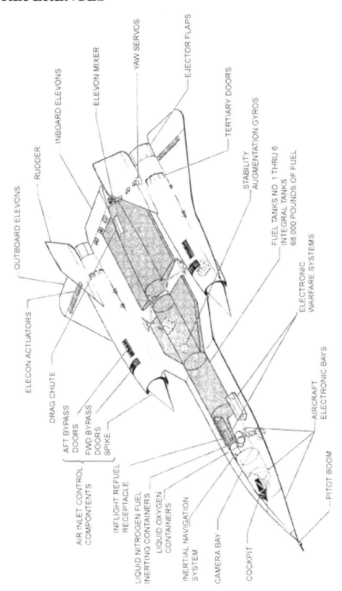

YAW SERVOS

EJECTOR FLAPS

ELEVON MIXER

INBOARD ELEVONS

TERTIARY DOORS

RUDDER

STABILITY AUGMENTATION GYROS

OUTBOARD ELEVONS

FUEL TANKS NO 1 THRU 6 INTEGRAL TANKS 68 000 POUNDS OF FUEL

ELECTRONIC WARFARE SYSTEMS

ELECON ACTUATORS

DRAG CHUTE

AFT BYPASS DOORS

FWD BYPASS DOORS

SPIKE

AIRCRAFT ELECTRONIC BAYS

AIR INLET CONTROL COMPONTENTS

INFLIGHT REFUEL RECEPTACLE

LIQUID NITROGEN FUEL INERTING CONTAINERS

LIQUID OXYGEN CONTAINERS

INERTIAL NAVIGATION SYSTEM

PITOT BOOM

CAMERA BAY

COCKPIT

Inventory of A-12s

Aircraft Number	Serial Number	Configuration	Number of Flights	Number of Hours Flown	Disposition
121	60-6924	Flight testing	332	418.2	Blackbird Airpark, Palmdale, CA
122	60-6925	Systems and flight testing	122	177.9	USS Intrepid Sea-Air-Space Museum, New York, NY
123	60-6926	Operations	79	135.3	Crashed 24 May 1963
124	60-6927	Training	614	1076.4	California Science Center, Los Angeles, CA
125	60-6928	Operations	202	334.9	Crashed 5 January 1967
126	60-6929	Operations	105	169.2	Crashed 28 December 1965
127	60-6930	Operations	258	499.2	Alabama Space and Rocket Center, Huntsville, AL
128	60-6931	Operations	232	453.0	CIA Headquarters, Langley, VA
129	60-6932	Operations	268	409.9	Crashed 4 June 1968
130	60-6933	Operations	217	406.3	San Diego Aerospace Museum, San Diego, CA
131	60-6937*	Operations	177	345.8	Southern Museum of Flight, Birmingham, AL
132	60-6938	Operations	197	369.9	USS Alabama Battleship Memorial Park, Mobile, AL
133	60-6939	Operations	10	8.3	Crashed 9 July 1964
134	60-6940	Drone operations	80	123.9	Museum of Flight, Seattle, WA
135	60-6941	Drone operations	95	152.7	Crashed 30 July 1966

*Numbers 6934 to 6936 were used for the Air Force's YF-12A fighter-interceptor version.

102

BLACK SHIELD Missions

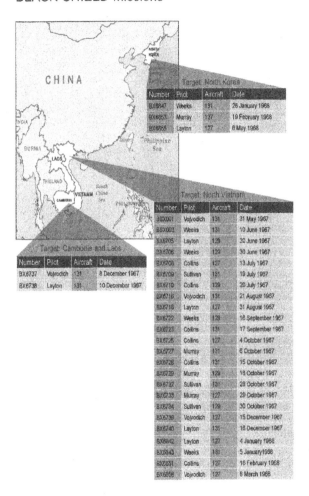

Target: North Korea			
Number	Pilot	Aircraft	Date
BX6847	Weeks	131	26 January 1968
BX6853	Murray	127	19 February 1968
BX6858	Layton	127	8 May 1968

Target: Cambodia and Laos			
Number	Pilot	Aircraft	Date
BX6737	Vojvodich	131	8 December 1967
BX6738	Layton	131	10 December 1967

Target: North Vietnam			
Number	Pilot	Aircraft	Date
BGX001	Vojvodich	131	31 May 1967
BGX002	Weeks	131	10 June 1967
BX6705	Layton	129	20 June 1967
BX6706	Weeks	129	30 June 1967
BX6708	Collins	127	13 July 1967
BX6709	Sullivan	131	19 July 1967
BX6710	Collins	129	20 July 1967
BX6718	Vojvodich	131	21 August 1967
BX6719	Layton	127	31 August 1967
BX6722	Weeks	129	16 September 1967
BX6723	Collins	131	17 September 1967
BX6725	Collins	127	4 October 1967
BX6727	Murray	131	6 October 1967
BX6728	Collins	131	15 October 1967
BX6729	Murray	129	18 October 1967
BX6732	Sullivan	131	28 October 1967
BX6733	Murray	127	29 October 1967
BX6734	Sullivan	129	30 October 1967
BX6739	Vojvodich	127	15 December 1967
BX6740	Layton	131	16 December 1967
BX6842	Layton	127	4 January 1968
BX6843	Weeks	131	5 January 1968
BX6851	Collins	127	16 February 1968
BX6858	Vojvodich	127	8 March 1968

Timeline of OXCART Milestones

1 July
First design studies for U-2 successor begin

29 August
Joint CIA/Air Force panel selects Lockheed's A-12 design

26 February
First A-12 completed, leaves Skunk Works for test site

25 April
First test flight

30 April
First official flight

4 May
First supersonic flight

6 November
First flight at Mach 2

21 April
Lockheed begins design studies for Mach 3 aircraft

11 February
CIA and Lockheed sign contract for 12 A-12s

15 January
First flight using two J58 engines

20 July
First flight at Mach 3

7 August
First flight of YF-12A variant

3 February
First sustained flight at design conditions (March 3.2 at 83,000 feet for 10 minutes)

29 February
President Johnson publicly announces existence of OXCART program

25 March
Last A-12 completed at Skunk Works

29 October
First SR-71 arrives at test site

22 December
First flights of SR-71 and M-21 carrying D-21 drone

27 January
A-12 flies for one hour and 40 minutes above Mach 3.1 for distance of 3,000 miles

12 November
CIA declares A-12 operationally ready

20 November
Maximum speed and altitude reached by one A-12 (Mach 3.29 at 90,000 feet)

5 March
First launch of D-21 drone from M-21

30 July
D-21 collides with M-21 just after launch, causing death of one crewman and leading to termination of program

28 December
President Johnson decides to end A-12 program by January 1968 (later extended to July)

5 January
Pilot Walter Ray is killed after ejecting from A-12

23 May
First A-12 flight across Pacific

31 May
First BLACK SHIELD mission over North Vietnam

26 October–3 November
A-12/SR-71 flyoff

5 January
Air Force cancels YF-12A program

26 January
First overflight of North Korea; Pueblo located

5 February
Air Force directs Lockheed to destroy tooling for all OXCART aircraft

21 March
First SR-71 mission over North Vietnam

6 May
Last (29th) A-12 mission, over North Korea

21 May
President Johnson reaffirms earlier decision to end A-12 program

4 June
Pilot Jack Weeks dies when A-12 crashes in checkout flight

21 June
Last A-12 flight, from test site to storage facility in California

26 June
CIA Intelligence Star awarded to BLACK SHIELD pilots

15 July
TAGBOARD program discontinued

1957 1958 1959 1960 1961 1962 1963 1964 1965 1966 1967 1968 1969 1970 1971

104

The OXCART "Family"

The A-12's unique design and characteristics became the foundation for three other versions of supersonic aircraft that Lockheed built for CIA and the Air Force: the YF-12A, the M-21, and the SR-71.

KEDLOCK: The YF-12A

In October 1962, the Air Force ordered three interceptor variants to replace the cancelled F-108A Rapier. The modified A-12, first designated the AF-12 and then the YF-12A, was designed and built under a project codenamed KEDLOCK. The aircraft's mission was to intercept new

Soviet supersonic bombers long before they reached the United States. It carried three air-to-air missiles and a second crewman who worked the fire control system. The Air Force initially envisioned a fleet of as many as 100, but only three were built and delivered during 1963-64. Secretary of Defense Robert McNamara cancelled KEDLOCK in early 1968 as a cost-cutting measure, and the aircraft were never deployed operationally. CIA was involved with the project only in giving up three A-12 airframes and helping write "black" contracts. The Air Force bore all the costs of the YF-12A, which was superseded by the F-111. Two of the aircraft were given to the National Aeronautics and Space Administration for research, and one was converted into a trainer for the SR-71 program.

TAGBOARD: The M-21 and D-21

In October 1962, CIA authorized the Skunk Works to study the feasibility of modifying the A-12 to carry and deploy a reconnaissance drone for unmanned overflight of denied areas. The project was codenamed TAGBOARD. The mother ship, redesignated the M-21 to avoid confusion with the A-12, was fitted with a second seat for a launch control officer (LCO) for the drone, called the D-21. It was 43 feet long, weighed over five tons, had a ramjet engine, could reach a speed of over Mach 3.3 at 90,000 feet, fly over 3,000 miles, and had the smallest RCS of anything Lockheed had yet designed. The drones would be launched well away from targets, fly their missions, and return to a preprogrammed location in international waters. There they would jettison a payload that a C-130 would snag in midair, and then self-destruct with a barometrically activated explosive device. In June 1963, the Air Force took over the project because it had overall charge of unmanned reconnaissance aircraft. Lockheed eventually built two M-21s and 38 drones, and its test pilot Bill Park flew all the M-21 flights. On the fourth TAGBOARD test on 30 July 1966, a launch mishap caused the mother ship to crash, killing LCO Ray Torick and prompting Kelly Johnson to end the program. Afterward the Air Force used B-52s to launch the drones against Communist Chinese targets in a project called SENIOR BOWL. Four missions were flown starting in November 1969. None was completely successful, and SENIOR BOWL was cancelled in July 1971.

SR-71 Blackbird

The best known version of the A-12 (above right) is the SR-71 Blackbird (above left), whose nickname has become eponymous with the entire set of OXCART variants. In December 1962, the Air Force ordered six "reconnaissance/strike" aircraft for high-speed, high-altitude flights over hostile territory after a nuclear attack, hence its original designator RS. Compared to the A-12, the SR-71 was about six feet longer, weighed 15,000 pounds more fully loaded, had more prominent nose and body chines and a two-seat cockpit, and carried additional optical and radar imagery systems and ELINT sensors in interchangeable noses.

With the added weight, the aircraft flew slower and lower than the A-12 or the YF-12A, but it carried more fuel and had a longer range. After an initial contract for six RS-71s, the Air Force ordered 25 more in August 1963. When President Johnson disclosed the aircraft's existence in July

1964, he mistakenly transposed the designator letters. Air Force officials let the error stand and came up with the Strategic Reconnaissance (SR) category instead. The fleet, based in the United Kingdom, Okinawa, and California, flew over 3,500 sorties from March 1968 until November 1989, when it was deactivated. In September 1994 Congress allocated funds to reactivate three SR-71s. Two aircraft and crews became operational during 1995 and 1996. In October 1997, President Bill Clinton vetoed further funding, and in June 1999 the SR-71 program was shut down again.

BIBLIOGRAPHY

Published Sources

Bissell, Richard M., Jr., with Jonathan E. Lewis and Frances T. Pudlo. *Reflections of a ColdWarrior: From Yalta to the Bay of Pigs.* New Haven, CT: Yale University Press, 1996.

Brown, William H. "J58/SR-71 Propulsion Integration," *Studies in Intelligence* 26:2 (Summer 1982), 15-23.

Crickmore, Paul F. *Lockheed SR-71: The Secret Missions Exposed.* London: Osprey, 1996.

Drendel, Lou. *SR-71 Blackbird In Action.* Carrollton, TX: Squadron/Signal Publications, 1982.

Goodall, James. *SR-71 Blackbird.* Carrollton, TX: Squadron/Signal Publications, 1995.

Goodall, James and Jay Miller. *Lockheed's SR-71 "Blackbird" Family: A-12, F-12, M-21, D-21, SR-71.* Hinckley, UK: Midland Publishing, 2002.

Graham, Richard H. *SR-71 Blackbird: Stories, Tales, and Legends.* St. Paul, MN: Zenith Press, 2002.

————. *SR-71 Revealed: The Inside Story.* Osceola, WI: Motorbooks International Publishing, 1996.

Haines, Gerald K. "The CIA's Role in the Study of UFOs, 1947-90," *Studies in Intelligence* 41:1 (1997), 67-84.

Helms, Richard with William Hood. *A Look Over My Shoulder: A Life in the Central Intelligence Agency.* New York: Random House, 2003.

Jenkins, Dennis R. *Lockheed SR-71/YF-12 Blackbirds.* North Branch, MN: Specialty Press, 1997.

Johnson, Clarence L. "Development of the Lockheed SR-71 Blackbird," *Studies in Intelligence* 26:2 (Summer 1982), 3-14.

Johnson, Clarence L. "Kelly" with Maggie Smith. *Kelly: More Than My Share of It All.* Washington, DC: Smithsonian Institution Press, 1985.

Landis, Tony R. *Lockheed Blackbird Family: A-12, YF-12, D-21/M-21 and SR-71 Photo Scrapbook.* North Branch, MN: Speciality Press, 2010.

McIninch, Thomas P. "The OXCART Story," *Studies in Intelligence* 15:1 (Winter 1971), 1-25.

Merlin, Peter. *From Archangel to Senior Crown: Design and Development of the Blackbird*. Reston, VA: American Institute of Aeronautics and Astronautics, 2008.

Miller, Jay. *Lockheed Martin's Skunk Works*. Leicester, UK: Midland Publishing, 1995.

Pedlow, Gregory W. and Donald E. Welzenbach. *The Central Intelligence Agency and Overhead Reconnaissance: The U-2 and OXCART Programs, 1954-1974*. Washington: Central Intelligence Agency, 1992. Chapter 6 on OXCART declassified October 2004.

Remak, Jeannette and Joseph Ventolo, Jr. *A-12 Blackbird Declassified*. St. Paul, MN: MBI Publishing Co., 2001.

————. *The Archangel and the OXCART: The Lockheed A-12 Blackbirds and the Dawning of Mach III Reconnaissance*. Bloomington, IN: Trafford Publishing, Co., 2008.

Rich, Ben R. and Leo Janos. *Skunk Works: A Personal Memoir of My Years at Lockheed*. Boston: Little, Brown, 1994.

Richelson, Jeffrey T. *The Wizards of Langley: Inside the CIA's Directorate of Science and Technology*. Boulder, CO: Westview Press, 2001.

Suhler, Paul A. *From Rainbow to GUSTO: Stealth and the Design of the Lockheed Blackbird.* Reston, VA: American Institute of Aeronautics and Astronautics, 2009.

Sweetman, Bill. *Lockheed Stealth.* St. Paul, MN: MBI Publishing, 2001.

Wheelon, Albert D. "And the Truth Shall Keep You Free: Recollections by the First Deputy Director of Science and Technology," *Studies in Intelligence* 39:1 (Spring 1995), 73-78.

Whittenbury, John R. "From Archangel to OXCART: Design Evolution of the Lockheed A-12, First of the Blackbirds." PowerPoint presentation, August 2007.

Wings of Fame. London: Aerospace Publishing, 1997.

Reference Documents

Central Intelligence Agency, Office of Special Activities. "Chronology, 1954-68." Declassified, June 2003.

Johnson, Clarence L. "Archangel Log." Undated.

————. "History of the OXCART Program." Burbank, CA: Lockheed Aircraft Corporation, 1 July 1968. Declassified, August 2007.

Web Sites

http://blackbirds.net

http://roadrunnerinternationale.com

http://www.habu.org

http://www.lockheedmartin.com

Made in the USA
Coppell, TX
13 November 2021

65687106R00069